**Quality Planning and Assurance:
Principles, Approaches, and Methods
for Product and Service Development**

# Quality Planning and Assurance

Principles, Approaches, and Methods for Product and
Service Development

*Herman Tang*
*Eastern Michigan University*
*Michigan, US*

*Registered Office*
John Wiley & Sons, Inc., 111 River Street, Hoboken, NJ 07030, USA

*Editorial Office*
111 River Street, Hoboken, NJ 07030, USA

For details of our global editorial offices, customer services, and more information about Wiley products visit us at www.wiley.com.

Wiley also publishes its books in a variety of electronic formats and by print-on-demand. Some content that appears in standard print versions of this book may not be available in other formats.

*Library of Congress Cataloging-in-Publication Data*
Names: Tang, He (Herman), author.
Title: Quality planning and assurance : principles, approaches, and methods for product and service
  development / Herman Tang.
Description: Hoboken, NJ : John Wiley & Sons, 2022. | Includes bibliographical references and index.
Identifiers: LCCN 2021024445 (print) | LCCN 2021024446 (ebook) | ISBN 9781119819271 (hardback) |
  ISBN 9781119819288 (pdf) | ISBN 9781119819295 (epub) | ISBN 9781119819301 (ebook)
Subjects: LCSH: Quality control.
Classification: LCC TS156 .T355 2022  (print) | LCC TS156  (ebook) | DDC 658.5/62--dc23
LC record available at https://lccn.loc.gov/2021024445
LC ebook record available at https://lccn.loc.gov/2021024446

Cover image: © nadla/Getty
Cover design by Wiley

Set in 9.5/12.5 pt and STIXTwoText by Integra Software Services, Pondicherry, India

10  9  8  7  6  5  4  3  2  1

# Contents

# Foreword

After graduating from the University of California, Santa Barbara, and the University of Michigan, I served as a faculty member and taught quality-related courses for over 40 years. I have consulted for Fortune 500 and smaller companies, the US State Department, and the US Army. Author of over 50 published papers and book chapters, I also directed more than 100 student graduate theses with nearly all research done in industry on quality-related problems.

For the retail consumer or casual observer, quality planning might seem a highly technical and completely boring topic with little relevance to the hyper-speed changes in the global pandemic-driven economy of 2021. However, in the view of this author, the "on-demand" and "customer-centric" economy and our expectations of online shopping and immediate delivery fueled by highly effective global supply chains has its antecedents going back at least to the ground-breaking published work on quality planning by Joseph Juran in the 1960s.

With little competition in the post-WWII boom, US monopolies and oligopolies "pushed" what they had the expertise to produce onto customers who had little choice in the marketplace. In the 1950s, the vacuum cleaner was a "Hoover," a refrigerator was a "Frigidaire," there were three television networks, the "Big Three" auto manufacturers in Detroit, and one telephone company. The recession of the 1970s triggered by OPEC's monopoly breaking "fuel crisis" ended the post-war boom and drove a change, at least in the US, to a highly competitive globally integrated and consumer-focused economy. Consumer spending in the US now constitutes more than 70% of the economy and nearly half the value of imports. Far-flung supply chains and intense competition require quality systems that result in few defects, poor product designs, or incorrect specifications – all this at breakneck speed!

With limited competition, automotive manufacturers and other monopoly/oligopoly goods producers could employ in-line attribute gauges that (mostly) sorted good from bad parts. However, the old "make and check" systems acceptable through WWII no longer suit. Taking quality "upstream" via quality planning has become the norm. Zero defects and Six Sigma goals replaced "acceptable levels of defects" enshrined in Mil Spec and other dated quality standards. The move to online shopping will further expand consumer choice, and diversify suppliers and online clearing houses – more choice, more consumers, and more producers in more locations mean greater need for fast and effective quality planning enabling rapidly changing products and services may be "pulled." Consumer-to-manufacturer (C2M) is only another iteration in the global economy's evolution requiring effective quality planning. Further, in a litigious society such as the US, product liability and other legal issues force producers to create and maintain a "paper trail" of planning documents. International, regional, and national quality systems compliance usually mandate specific procedures of quality planning.

Dr. Herman Tang's work reflects his many years of experience as an engineer, quality specialist, researcher, and professor. This book is no mere random arrangement of topics thrown together to create a textbook and has great value for university students as well as novice and experienced quality professionals. The sequence of material is logical, well-documented, and accessible. The book is equally valuable as a text and a reference. Readers should keep this book handy on their shelf of prized reference sources, and consider it as an authoritative guide to quality planning specifically and the practice of quality systems in general.

*Dr. Walter W. Tucker, PhD*
Professor Emeritus, Eastern Michigan University

# Foreword

I am a Fellow of the American Society for Quality (ASQ) and Chief Expert of ASQ Shanghai LMC. I had worked in quality at Ford Motor Company for 28 years before my retirement in 2020. At Ford, I held various technical and leadership positions, including Vice President of Ford China Quality and New Model Program Launch, Vice President of Ford Asia Pacific Quality and New Model Program Launch, Ford Corporate Executive Technical Leader, and Member of the Technology Advisory Board. It is my honor to write a Foreword for this book.

According to Dr. Joseph Juran, a quality management system has three cornerstones: quality planning, quality control, and continuous improvement. Quality planning acts as a fundamental and proactive function to the other two cornerstones. I believe we can all do even better jobs if we devote more efforts and resources to quality planning. There is a need for a comprehensive textbook and reference book that captures quality planning subjects. I appreciate that this book intends to do just that.

I have reviewed all the chapters of Dr. Tang's manuscript. I have found that it includes all the essential elements for quality planning practices, reviews many approaches, and presents the contents systematically and logically. I appreciate the book's broad coverage on principles and applications, with numerous diverse examples, not limited to a particular industry or discipline. The broad coverage and various examples make this book suitable for graduate students in almost all majors and practitioners in various fields, such as product development, manufacturing systems, and service development.

I furthermore like that Dr. Tang emphasizes innovative and critical thinking in quality work, which is crucial but often gets ignored in quality learning and practice. Quality is both a science and an art; a quality project can be unique in different situations, and remains case dependent. In my judgment, being guided by this book, readers can learn different aspects and practices, creatively make work effective, and develop new quality methods.

Based on my experience and review, I strongly recommend this comprehensive book to quality students, professionals, and managers. Through practice and referring to this book, readers would develop their own quality planning expertise, and contribute to their workplace and the professional community at large.

*Jay Zhou, Ph.D.*
ASQ Fellow, Vice President of Ford China Quality and
New Model Program (ret.)

# Preface

## Book Intent

### Principles Focused

Quality planning has been a fundamental industrial practice for several decades, yet there are few comprehensive quality planning textbooks dedicated to the understanding of this subject at the undergraduate and graduate levels. In the quality field, professionals often consider Toyota as a role model for best practice. While one can learn the principles of Toyota quality, its specific practices are not necessarily applicable for every situation. In *The Toyota Way to Service Excellence*, Dr. Liker and Ross stated, "the Toyota Way training was designed to teach principles rather than specific methodology" (p. 32). Similarly, this book focuses on the fundamental principles of quality planning, and extrapolates on their applications in various industries throughout each chapter.

As current and future quality professionals, you can start learning these principles, with supporting application examples in this book, and later apply them to your unique applications. Like one of my students said, "What I enjoyed most about this course was taking the information learned in this course and being able to utilize it within the industry that I currently work in."

### Broad Applications

Much of the quality literature focuses on physical and tangible goods (called products in this book). A primary reason might be that the prominent quality figures and their practices, such as Deming's 14 Points, Taguchi loss function, Juran trilogy, Kano model, Hoshin planning, and Toyota Production System, were all created for products and manufacturing.

Because of a vast number of different types of business, theories, and applications, it would be impossible to cover all of them in a single volume. Therefore, this book introduces main topics in quality planning based on the common practices in industry, and extends a discussion to other types of fields as a unique feature. For example, the automotive industry started its systematic development and implementation of quality planning practice a few decades ago. These principles and approaches to quality planning have been recently adopted by other sectors, such as healthcare.

An effective way to conduct quality work is to learn from other professionals. This book collects and integrates the good practices across industries by citing over 370 scholarly papers and other types of sources in various areas. The cited research and development literature are recent and provide the latest advances in the fields in addition to demonstrating the fundamental principles of quality planning. You may even find new opportunities throughout this book to contribute in furthering the quality planning field.

## Book Content

### Overall Flow

The contents of the book are arranged from general principles and approaches to specific methods and tools in eight chapters, as shown in the figure. To learn the subject systematically, it would be a good idea to follow these chapters in order. Note that Chapters 2 and 3 could be studied in parallel, as the tasks in these chapters may be planned and executed at the same time. Similar consideration can be taken for the two pairs of Chapters 4 and 5, and Chapters 6 and 7.

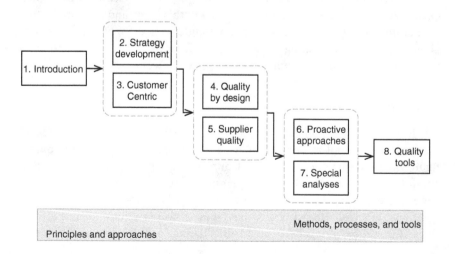

Quality management systems and quality planning are both sciences and art forms. In quality practice and study, one can make their work of integrating the principles, approaches, and methods more effective through firm comprehension, planning, and implementation. This book integrates approaches, methods, etc. into each of their fundamental principles, along with copious amounts of diagrams and tables, for more effective and comprehensive learning.

## Eight Chapters

Chapter 1 Introduction to Quality Planning. The first chapter starts with the concepts, meanings, and dimensions of quality. The chapter introduces a quality system and the planning role in a system, and reviews the processes and general guidelines of quality planning.

Chapter 2 Strategy Development for Quality. This chapter presents the overall process of policy management. It discusses Hoshin X-matrix development and considerations in policy development. The chapter also addresses risk management and discusses pull and push strategies.

Chapter 3 Customer-centric Planning. This chapter reviews the concerns of fundamental planning. First, the chapter discusses the characteristics of products and processes. The chapter also introduces the categories of quality and the Kano model, from customer perspectives. Then, the chapter explains the principle and process of quality fuction deployment (QFD) development. Lastly, the chapter introduces the relatively new subject of affective engineering (AE) in quality planning.

Chapter 4 Quality Assurance by Design. This chapter discusses design verification and validation, starting with the general process of design reviews, tools such as design review based on failure mode (DRBFM), and the concept of concurrent engineering (CE) into product and service development. The chapter also addresses the variation characteristics and their influences in design.

Chapter 5 Proactive Approaches: Failure Modes and Effects Analysis and Control Plan. This chapter presents the concepts, development processes, and considerations of Failure Modes and Effects Analysis (FMEA) and Control Plan, which can be used in proactive quality planning for product and service development.

Chapter 6 Supplier Quality Management and Production Part Approval Process. This chapter introduces the production part approval process (PPAP), originally developed in the automotive industry, and reviews its applications, key points, and other standards in the field of supplier quality assurance. This subject is vital, as suppliers are an integral part of business development and operation.

Chapter 7 Special Analyses and Processes. This chapter discusses four special processes: measurement system analysis (MSA), process capability study, design change management, and quality auditing. These processes are essential to quality assurance, particularly for the planning and development phases of a product or service.

Chapter 8 Quality Management Tools. The last chapter presents and compares several problem-solving and continuous improvement processes, such as DMAIC, 8D, and PDCA. Then, the chapter reviews 14 conventional tools for quality assurance during product and service development and execution. These processes and tools are widely implemented in quality management.

## Learning Exercises

Quality planning is not only about textbook principles and methods, but also their practices and implementations. Exercises are vital to learning and mastering these principles and methods. Two types of exercises are developed: 20 Review Questions and 10 Mini-project Topics at the end of each chapter.

Both types of exercises complement individual thinking and class team-learning activities. The Review Questions are for an immediate classroom/online discussion, anticipating quick answers. The Mini-project Topics are more in-depth, and require more effort on materials research, critical thinking, and short essay writing. Most Review Questions and Mini-project Topics are exchangeable. A topic from the Review Questions can be expanded to a Mini-project, and conversely, a Mini-project Topic can be simplified for immediate discussion.

The instructional materials of this text are available for teaching, including a sample course syllabus, project development guidelines, class exercise instructions, and discussion instructions. These instructional materials can be direct references for instructors and students, in a classroom setting or for online learning. Instructors who use this text can also send the author requests for instructional supplemental materials.

# Acknowledgments

## Teamwork

Special thanks go to Dr. Walter Tucker, a professor emeritus at Eastern Michigan University, who had taught this subject for over 30 years. His mentorship on quality pedagogy is essential to this text. Special appreciation to Dr. Jay Zhou (ASQ Fellow and retired Asia Pacific Vice President for quality at Ford Motor Company), Dr. Christopher Kluse (Professor at Bowling Green State University), and Carlos Zaniolo (Manager at Volvo Group), who reviewed all the chapters of the manuscript and offered suggestions for improvements.

I also extend my thanks to senior professionals in industries and academia: Santos Aloyo (Becton Dickinson), Marc Deluca (Ford Motor Company), Jon Gawlak (SGS North America), Ryan Gingras (Pratt & Whitney Autoair), Qian Harris, Bryan Jakubiec (Magna International), Mike Smith, and Dr. David Tao (University of Michigan), who provided their comments and suggestions. Eastern Michigan University student Brendan Ostrom helped with final manuscript proofreading.

I am also grateful to the five anonymous reviewers who provided constructive comments on this book's proposal and draft chapters, as well as Wiley's acquisition, project, editing, and publication teams who played integral roles in the quality publication of this book.

I would also like to thank Eastern Michigan University for supporting this book manuscript's preparation with the 2021 Faculty Research Fellowship Award. While teaching this subject at EMU, I really enjoy working with the students, many of whom are experienced professionals in quality fields across many industries, with insights and experience that have helped expand the applications and scope of the manuscript. I appreciate the authors and organizations for their works that are cited in the book.

Finally, yet importantly, I would like to thank my family; my wife for her understanding and full support, and our sons Boyang and Haoyang for their help and advice in bringing this volume to fruition.

## Your Feedback

Quality planning is a science and an art, one of the most broad and diversified practices. I have put a lot of thought into the manuscript that reflects my personal understanding and experience in this realm. I hope this text provides a useful reference for learning, practice, and advancement of quality planning principles and approaches.

Reader's insights, comments, and feedback are welcomed and appreciated, to help continuously improve this text. Please send your comments, criticism, and suggestions to htang369@yahoo.com and htang2@emich.edu. I will carefully review them for a future edition of this book. I wish you the best of success in your quality professional work.

*He (Herman) Tang*
*Ann Arbor, MI, USA*
*March 2021*

# About the Author

Dr. He (Herman) Tang is an associate professor with the School of Engineering at Eastern Michigan University. His experiences and interests concern the fields of mechanical, manufacturing, and quality engineering, among others. He has taught nine graduate courses in quality, and has been responsible for the Master of Science in Quality Management program at Eastern Michigan University for several years. He has served as an associate editor and reviewer for several scholarly journals and conferences, and as a panelist for the National Science Foundation. Dr. Tang has  published four technical books, two book chapters, and many scholarly journal papers, and has delivered invited presentations. His previous book is *Engineering Research: Design, Methods, and Publication* published by Wiley in 2020. Dr. Tang earned his doctoral degree of Mechanical Engineering from the University of Michigan–Ann Arbor, master's degree and bachelor's degree of Mechanical Engineering from Tianjin University, and MBA in Industrial Management from Baker College.

# 1

# Introduction to Quality Planning

## 1.1 Quality Definitions

### 1.1.1 Meaning of Quality

**Definition of Quality**

*Quality* is a common word both in the workplace and at home, so it might be interesting to ask exactly what the word quality means. There are certainly numerous definitions of quality. Regarding this ambiguity, the American Society for Quality (ASQ) states that quality is "a subjective term for which each person or sector has its own definition. In technical usage, quality can have two meanings: 1) the characteristics of a product or service that bear on its ability to satisfy stated or implied needs; 2) a product or service free of deficiencies" (ASQ, n.d. a).

Several quality pioneers established the foundation of present quality definitions. For example, Dr. Joseph Juran viewed quality as a fitness for use and leading to customer satisfaction (ASQ n.d. a). Philip Crosby thought of quality as a conformance to requirements (Crosby 1979). Formally, ISO 9000:2015 defines quality as the degree to which a set of inherent characteristics fulfills requirements (ISO 2015).

While ambiguous on the surface, defining quality in a given professional field can be an intriguing exercise, which helps us think about its importance as a target, a process, and a system when applying quality principles and approaches to real-world situations. Here are several examples of quality definitions, each of which may reflect aspects to the different fields and view angles of quality practitioners:

"Fit for purpose while robust enough to uphold product/service integrity and value" (Glodowski 2019). This definition addresses the robustness to fit for purpose.

"Quality is the art of always pleasing the customer while ensuring your bottom line is met" (Toure 2019). This definition recognizes quality is the art of meeting a customer's need.

*Quality Planning and Assurance: Principles, Approaches, and Methods for Product and Service Development*, First Edition. Herman Tang.
© 2022 John Wiley & Sons, Inc. Published 2022 by John Wiley & Sons, Inc.

"Sincere and considerate actions taken to either fulfill or exceed user expectations" (Mori 2018). This definition focuses on implementation processes and actions.

"The required satisfaction provided by a good or service as expected or imagined by the customer" (Stevens 2018). This definition also touches on an important aspect of the perceived service of quality to customers.

Considering the many aspects of quality illustrated in just these four examples, one may find that it can be difficult to encompass the entirety of such a broad concept in a single sentence. To help with this, one may use a few keywords to understand the general meaning of quality when viewed in relation to industry, goods, and services:

- Quality is <u>customer</u> (or end-user) oriented.
- Quality is a distinctive characteristic or degree of <u>excellence</u> of something.
- Quality is adherence to <u>specifications</u> (or standards/regulations) by a product or service.
- Quality is a summary description of multiple <u>dimensions</u> and aspects.

Readers may have even more keywords to add to this list, simply based on their own experience and insights. In addition, these keywords may have different weights or significance when one addresses unique situations, products, or problems. While the definition of quality can be subjective to person and place, the broader concept is objective as a field of study, and based on the same basic truths about how quality can be defined.

**Measurement of Quality**
The aforementioned keywords, i.e. customer demands, design specifications, and quality dimensions, represent the references for quality measurement and analysis. As a foundation of quality management, quality measurement is an evaluation:

- Of a specific status or result from a product or service
- Of a process or system of processes involved with a product or service

The first action of a quality measurement process is to collect current data, shown in Figure 1.1. For example, the US Environemental Protection Agency (EPA) measures the quality of the air by collecting and analyzing the presence of specific compounds that cause pollution. Their measurement for reporting is the Air Quality Index (AQI) (EPA 2019). The AQI tells us how clean the air is, and what associated health effects might be a concern.

For the quality of a product, measurement and analysis are based on that product's design specifications. For example, software development has several functional specifications, e.g. technical details, data manipulation, and processing efficiency. The requirements, test processes, and criteria are predefined as a

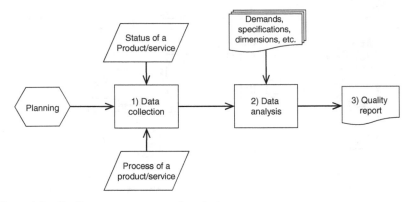

**Figure 1.1** Quality measurement and analysis process.

guideline or standard. Also, since the quality of a product or service has multiple dimensions, so too must the corresponding measurements and analyses of that quality, which will be discussed in detail later.

For the quality of a manufacturing or service operation, one may collect process information and output in terms of adherence to standards, design specifications, etc. of that given operation. Measuring output is mainly for determining the status and results of an operation. Measuring the process itself can provide insight into understanding how the output of an operation is generated, thereby helping to find reasons why the output meets (or fails) the requirements of the given input. Therefore, it is often important to measure both process and output.

Quality measurement reports are normally presented in terms of well-defined indexes. For example, a quality indicator of healthcare quality is called the Patient Safety Indicator, which shows avoidable safety events that represent opportunities for improvement in the delivery of care (AHRQ n.d.). Another similar one is called the Prevention Quality Indicator, which is used to identify the conditions in which good outpatient care can potentially prevent the need for hospitalization. Quality indicators are field specific, and often have multiple ones for a product and service. For example, over 30 quality indicators were considered in blood establishments at the international level (Vuk 2012). In addition to direct measurements on a product, process, or service, the quality variance, or the differences between the individual measurements, can be used as an indicator as well.

### 1.1.2 End-customer Centricity

#### Internal Customer

A customer is an individual or business entity who receives and uses a product or service. A large or complex business operation has various units, so their

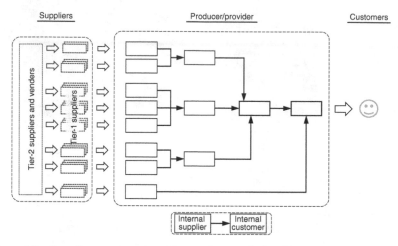

**Figure 1.2** Supplier–producer/provider–customer relationship.

working relationship of providing and receiving can be multifaceted, much like the network of a community. An example is shown in Figure 1.2. Suppliers provide parts and/or services to a product producer or service provider. The latter is the customer to the suppliers, while the producer/provider works with suppliers to provide a product or service to ultimate customers.

Inside an organization, there is also a customer–supplier relationship. An internal customer (or client) can be a person, operation, or group within an organization, who performs their jobs when receiving an output, part, and/or assistance from one or more internal persons, operations, or groups. For instance, in a surgical operating room, the ultimate customer is the patient, while the surgeons can be viewed as internal customers of the nurses and technicians who assist the surgeons.

Therefore, an operation in a business system can be viewed as the internal customer of all upstream or supporting operations, and at the same time is an internal supplier to downstream or other associated operations. If one analyzes the suppliers, inputs, process, outputs, and customers (SIPOC) of a business operation, they may know their relationships. Table 1.1 shows an example of a SIPOC analysis for vehicle manufacturing. In a SIPOC analysis, the suppliers and customers can be either internal or external.

The concept of an internal customer can be helpful to build an effective relationship between operations and teamwork for an organization's ultimate customers. In many cases, internal customers are less obvious than external customers because of management structure, lack of financial transaction, and/or complex organizational functions. For example, it can be difficult to define supplier–customer relationships for some departments in a matrix organization.

**Table 1.1** SIPOC of vehicle manufacturing operations.

| Operation | Supplier | Input | Process | Output | Customer |
|---|---|---|---|---|---|
| Part supplier | Part/material manufacturers | Raw materials, parts, etc. | Various | Components, parts, etc. | Vehicle assembly plants |
| Body shop | Part suppliers | Components, parts, materials | Joining, etc. | Framed car bodies (body-in-white) | Paint shop |
| Paint shop | Body shop, material suppliers | Body-in-white, materials | Painting, sealing, etc. | Painted car bodies | General assembly shop |
| General assembly shop | Paint shop, part suppliers | Painted bodies and components | Installation, etc. | Completed vehicles | Car buyers |

**Toward End-customer Satisfaction**

From the viewpoint of quality, the customer of a business is the ultimate end-user of a product or service. Internal suppliers and the relationship between internal suppliers and internal customers are the enablers that satisfy these ultimate end-users. Customer satisfaction as a goal is for ultimate customers, but may or may not be for internal customers. All internal customers and suppliers in a system work together to make their operations smooth and effective to collectively provide a good product or service to the external customers.

In an organization, it is possible that people are more concerned about their internal customers, which sounds parochial and may be disadvantageous to external customers. There can be a conflict of interests between internal customers and external customers sometimes. It is the senior management's responsibility to encourage and guide the internal supplier–customer teamwork for the sake of external customers.

Furthermore, external suppliers or vendors should treat the receivers of their products and services as customers. Altogether, a product or service provider and suppliers should define, plan, and implement collective work for the end-users, as illustrated in Figure 1.3. The producer–supplier partnership works as one team to build a mature trust, help each other, and grow together. Treatments on customer needs and supplier quality will be discussed in depth in Chapters 3 and 6, respectively.

Regard the familiar adage: "the customer is always right." It is important because happy customers are more likely to buy a product or service again from a company who meets or exceeds their needs. Adopting this adage, one also needs to understand the customer expectations and variation of processes, to be

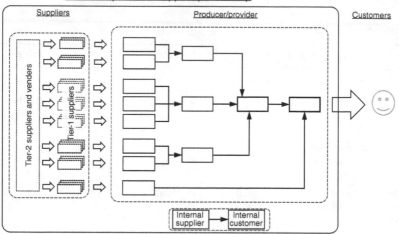

**Figure 1.3**  Producer/provider–supplier partnership for ultimate customers.

detailed in Chapter 4. With this familiarity, one may have accurate and realistic goals for customer satisfaction, quality planning, and execution.

Customer-centric quality planning and its associated processes take time, and their effects may or may not be immediately visible. Adopting quality-planning principles may also need a culture change for some organizations, particularly those with a well-established process, with a mentality of firefighting problem-solving, and/or with a focus on cost reduction. In such cases, changing the primary motivation to customer satisfaction can meet some resistance.

### 1.1.3  Dimensions of Product and Service Quality

**Product Quality**

One can view quality in several aspects. For a product, one may view its quality in performance, features, reliability, aesthetics, and so on. For example, the quality of a passenger vehicle has eight dimensions (Tang 2017), as shown in Table 1.2.

Depending on the type of product, there can be different dimensions or characteristics, such as ergonomic and environmental performances. For example, the serviceability of cellular phones can be ignored, as a broken cellular phone is normally replaced rather than repaired. While the serviceability of automobiles and air conditioners has significant impacts on the after-sales service cost and customer satisfaction (Syahrial et al. 2019). For software products, the

**Table 1.2** Quality dimensions of passenger vehicles.

| Dimension | Description | Example |
|---|---|---|
| Performance | Primary operating characteristics | Acceleration: 8.2 s for 0–60 mph |
| Safety | Crashworthiness and crash avoidance (performance) | 5-star rating of crash tests by NHTSA |
| Features | Secondary performance characteristics | Folding seats and DVD/TV/ Bluetooth function |
| Reliability | Probability of working consistently well without major failure | Running three years without major issue |
| Durability | Measure of a product lifespan (replacement preferred over repair) | Engine 95% reliable (without major issues) in 3 years |
| Aesthetics | Based on looking, feeling, sound, etc. | Flaming red color (subjective) |
| Conformance | Meet established standards and expectations | No water leaks |
| Serviceability | All related to services, including cost, speed, service professionalism | Routine service from a dealer |

*Source*: Tang, H., (2017). *Automotive Vehicle Assembly Processes and Operations Management*, ISBN: 978-0-7680-8338-5, Warrendale, PA: SAE International.

quality model is standardized (ISO 2011) for the eight characteristics of software quality:

1. Functional suitability: Functional completeness, correctness, and appropriateness.
2. Reliability: Faultlessness, availability, fault tolerance, recoverability, and failsafe.
3. Performance efficiency: Time behavior, resource utilization, and capacity.
4. Operability: Understandability, learnability, user error protection, user interface aesthetics, and accessibility.
5. Security: Confidentiality, integrity, non-repudiation, accountability, and authenticity.
6. Compatibility: Co-existence and interoperability.
7. Maintainability: Modularity, reusability, analyzability, modifiability, and testability
8. Flexibility: Adaptability, scalability, installability, replaceability, and portability.

Furthermore, different aspects or dimensions are not equally important, as they depend on the consequence of poor quality of a product or service on average. For example, the safety of passenger vehicles and medical instruments is their most important consideration, thus they are seriously addressed by manufacturers, tightly regulated by the government, and largely expected by customers. The safety of laptops is important as well, but may be less concerning because it has a less serious impact on the safety of a passenger vehicle or medical instrument. Significance of quality dimensions is also related to the type of customers, e.g. age, gender, geographic location, etc. The importance levels of dimensions can be presented in a percentage contribution to the total quality of a product. Figure 1.4 shows an example. It is recommended that a weight for each dimension be developed before the design phases.

**Service Quality**
The attributes of service quality can be more subjective and more directly relating to a customer's feeling and perception than those of product quality do. In other words, service quality is about the direct relation between a customer's expectations and a provider's performance. An understanding of service quality (SQ) generally is

$$SQ = P - E$$

where P is perceived performance and E is perceived expectations (Lewis and Booms 1983). Note, SQ has multiple dimensions, as do P and E.

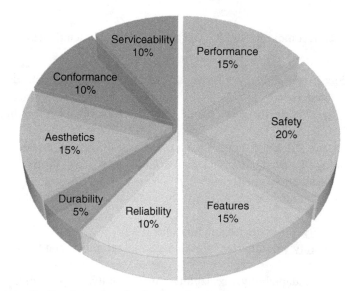

**Figure 1.4** Quality dimension distribution of a product.

From a consumer's perspective, service quality includes 10 aspects: reliability, responsiveness, competence, access, courtesy, communication, credibility, security, understanding, and tangibles (Parasuraman et al. 1985). The five main dimensions of service quality, sometimes called the SERVQUAL model, are briefly explained in Table 1.3. For a specific service, though, the dimensions can vary in a given situation. For example, the quality of a restaurant includes six dimensions: food quality, facility comfort, cleanliness, timeliness, aesthetic, and personnel service quality (Tuncer et al. 2020).

As with the dimensions of product quality, the dimensions of service quality are not equally important to one another, or they might carry different weights. For some services, such as personal and computer services, security is very important. In many other cases, responsiveness is a major factor affecting customer satisfaction, e.g. for premium casual restaurants (Saad et al. 2020). In other words, the main dimensions of service quality and their significance depend on the type of service. In addition, the rendering of products or services by a caring, friendly person often is a key to ensuring service quality dimensions.

A service is often associated with a product, either physical products (cars or computers) or nonmaterial products (bank accounts or loan products). In such cases, a product and service can be combined as a package, e.g. specialty software, for customers. On the service side, one quality model in software is called "Quality in Use" (ISO 2016). It has five characteristics to meet user's needs to

**Table 1.3** Main dimensions of service quality.

| Dimension | Description | Action |
| --- | --- | --- |
| Reliability | Ability and correctness as described and promised to perform a service accurately and consistently at the first time | Deliver at the designated time |
| Responsiveness | Promptness of willingness or readiness to customer needs | Quickly answer and resolve issues |
| Assurance | Capability to convey trust and confidence with customers (with competence, respect, communication, and attitude) | Build long-term relationship |
| Empathy | Understanding, caring, and genuine concern, paying individual attention with sincerity for customers | Learn and recognize various needs |
| Tangibility | Related to physical facilities and company aesthetics appealing to customers | Human representative to customer |

achieve specific goals: 1) effectiveness, 2) efficiency, 3) satisfaction, 4) freedom from risk, and 5) context coverage in use.

**Quality of Product–service Hybrid**

Customers often receive a product and related service as a combination or a product–service hybrid. For example, a wireless phone package includes a cell-phone and a service plan as a bundle. In such cases, customer's satisfaction is related to the quality of both product and service. Another example is food, which is a product, and includes some aspects of service. Food quality has nine dimensions (EC n.d.):

- Safety
- Ethical
- Sensory
- Nutrition
- Aesthetical
- Functional
- Convenience
- Authenticity
- Origin

Considering all the characteristics and dimensions of quality, one can see their relationships and contributions to customer satisfaction are complex. The connection between the quality of a product and the associated service is illustrated in Figure 1.5, in terms of common dimensions of quality.

In general, a good product is a foundation of good service, while other characteristics and dimensions influence one another. In planning for optimal quality, one needs to address not only the characteristics and dimensions of a product and service, but also their interactions and integration. The former has been extensively studied, while the latter can be a new research focus.

New technologies, such as artificial intelligence, in products and services, like the digital assistant, play an increasingly important role in customer satisfaction. Their significance and depth of impact on quality are active subjects of contemporary study (Brill et al. 2019).

### 1.1.4 Discussion of Service Quality

**Product vs. Service Quality**

Comparing the dimensions of product and service, one should consider the separation between unique characteristics of service quality from those of product quality. One attribute of a service is its intangibility. Many types of service provided have no solid proof. The intangibility is difficult to evaluate objectively. For example, one can call a service 800 number for a concern on a

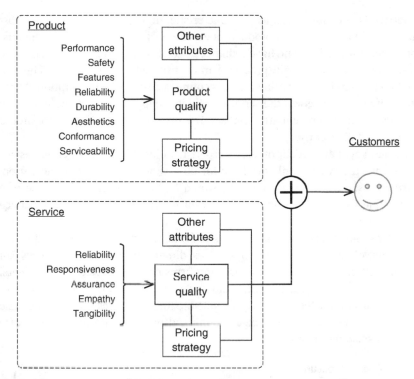

**Figure 1.5** Characteristics and dimensions of product and service quality.

defective product. The service representative may resolve the problem by refund, but one might be still unhappy because the representative takes too long or one needs the product now, and the refund does not fulfill the greater need of the customer. Intangibility is also a confounded aspect of service quality, as some other dimensions can be intangible or even invisible, depending on the type of service. Nowadays, more and more services go through the internet, which intensifies their intangibility.

Physical products can be made homogeneously, while a service is often heterogeneous or has a significant variability in each service event, largely because service is highly people driven. For example, in a bank, it is difficult to encapsulate the experience for every customer with a unified criterion. The service performance of different representatives are different, even the same representatives can vary from day to day and customer needs can be even more diverse. Thus, heterogeneity is another characteristic of service quality. A service system should accommodate such variations where anomaly is the norm. Besides, the perception of service quality from all customers is different. Because of these differences, many standards and processes for product quality cannot apply to service quality.

Service is demand driven as a characteristic, in a "pull" fashion that is discussed in Chapter 2. For a product, one can check its quality before the customer gets the product, and have a chance to get it right. A service is inseparable from a demand and typically produced and consumed simultaneously. The production of a service, e.g. online retail, is connected with its consumption. There are limited ways to assess the service while it is happening, and one may have to evaluate and improve it after the fact. Because of this, service quality should be managed differently.

The last key characteristic of service is perishability. While a product can last for some time, a service is only available for observation during a definite period. Further, a service capacity cannot be stored and carried forward for sale in a future time, as excess product might be sitting on a shelf or warehouse. A service cannot be returned or resold once it has been consumed. Therefore, a service provider should utilize its capacity to meet demands with given time windows.

Because of the massive variety of services, dimensions and characteristics can be very different. For example, consulting firms provide services to customers. The customers prioritize receiving knowledge and solutions that are useful in not only current business issues but applied to future business hurdles (Benazic and Varga 2018). Customers' expectations and priorities are unique and vary from one customer to another.

**Service Quality Measurement**

In addition to the dimensions and characteristics of service quality, measurements and metrics can also be different. Quality measurements can be objective, subjective, or anywhere in between, such as waiting time and patients' perspective satisfaction for the services in a hospital. The satisfaction of the waiting time is affected by other factors, such as a patient's feeling, mood, schedule, the time of a day, and weather. Besides, there can be a certain correlational relationship between objective measurements and perspective data.

To measure a customer's perspective viewpoints and feelings, a questionnaire survey is commonly used. A quick survey with a few questions can be conducted immediately after a service is delivered. Similarly, a follow-up survey of a large service can be conducted with 10–15 questions. Often, an open-ended comment window is also provided for customers to give detailed opinions that a rating question may not be able to capture.

An interesting investigation method of service quality is called mystery shopping. In these cases, an internal or external member of an organization as an undercover investigator examines service quality, based on a defined process. This approach can provide important information and independent observation. As a research approach, the applications of mystery shopping have been growing recently. Studying this approach on its effectiveness and reliability in different fields is also active (Minghetti and Celotto 2013, Blessing et al. 2019, Dutt et al. 2019).

A rating for surveys may be designed in a three-point or five-point (Likert) scale. If on a five-point scale, the levels may be: 1) very unsatisfied, 2) unsatisfied, 3) neutral, 4) satisfied, and 5) very satisfied. For example, one may use the five faces that range from a crying face to a huge smile. The data analysis of a large survey consists of four steps: 1) initial grouping, 2) category development, 3) subset analysis, and 4) thematic coding, which can be done by trained professionals using computer software.

Even though some quality methods, standards, and measurements of product quality may not directly apply to the quality of service, the principles are still good for services, if only applicable with specific focuses or criteria. In many cases, the term "product" can be replaced by "service" with necessary modifications on some approaches and methods. Service quality planning and execution should address the specific characteristics mentioned above.

## 1.2 Quality System

### 1.2.1 Quality Management System

**Meaning of QMS**

When applying quality principles, one must treat quality as an integral part of an entire business system, rather than a technique or department, to make quality planning and execution effective. From this point, quality is not a task but a business foundation and subsystem.

One can view quality management as a collection of quality theories and practices on concepts, principles, methods, and processes. ISO 9001 (ISO 2015) is the international standard that specifies the requirements to implement a quality management system (QMS), with a focus on processes and documentation. Similarly to ISO 9001, there are specific industry and region standards, such as VDA 6.x, that are the German automotive industry standards for QMS (VDA 2016). The effectiveness of these standards and their implementation are evolving, and are the subject of contemporary study, e.g. (Franceschini et al. 2016; Sun et al. 2019; Nurcahyo et al. 2021).

There are other definitions of a QMS, such as "a formalized system that documents processes, procedures, and responsibilities for achieving quality policies and objectives." (ASQ n.d. b) In practice, people may refer to a piece of software, which is used to manage routine quality tasks, as a QMS.

There are different standpoints on the definitive elements of a QMS from various organizations and researchers. The common elements and requirements for a QMS include:

- Quality objectives and policy
- Customer satisfaction focus

- Quality manuals
- Procedures, instructions, and records
- Quality control and assurance processes
- Quality data management
- Continuous improvement opportunities
- Quality analysis

The US FDA declares that a pharmaceutical quality system consists of four elements (FDA 2009):

- Process performance and product quality monitoring system
- Corrective action and preventive action (CAPA) system
- Change management system
- Management review of process performance and product quality

A study was conducted on the significance of the six elements of a QMS, based on a survey of 238 plants in the US, Japan, Italy, Sweden, Austria, Korea, Finland, and Germany (Zeng et al. 2013). The authors found that the six elements have close relationships to the quality management of customer companies with different standardized coefficients, shown in Figure 1.6.

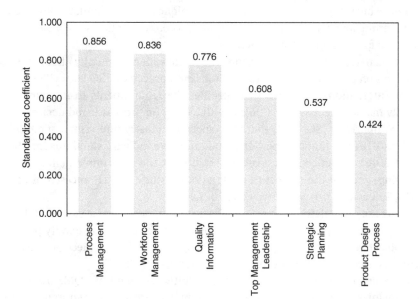

**Figure 1.6** Study on six elements of QMS.

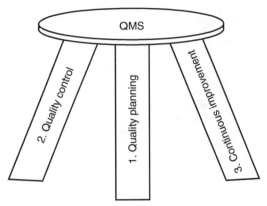

**Figure 1.7** Three pillars of QMS.

### Juran QMS Trilogy

An important visualization for a QMS is to picture three pillars or cornerstones: quality planning, quality control, and continuous improvement (see Figure 1.7). This concept is also known as the Juran trilogy (Juran 1986).

1. A quality journey begins with <u>quality planning</u>. The first task of quality planning is to identify customers and recognize their expectations. Based on the understanding of the voice of customers, one can develop a goal and plan for a product or service, and develop its features. The goal and plan should transfer the vision of an organization to the tasks of every member in the organization, such as managers, engineers, and operators.

Quality planning is a base for the most tasks and activities of quality management practice. The outcomes of quality planning largely affect the work and performance of the other two pillars, as illustrated in Figure 1.8. Good quality planning makes quality control and continuous improvement less challenging and more promising in terms of efforts and costs. Industry practices prove that good quality planning can make later operations (manufacturing, service, etc.) smoother and more cost effective.

2. The second pillar of a QMS is <u>quality control</u>. During the early realization phases of a product or service, one should inspect, monitor, and control the quality of operations and processes. The out-of-control data in the operations and processes need to be analyzed to find root causes for correction. A common practice is to use various statistical process control (SPC) tools to ensure realization operations and processes in control.

3. After an operation and process is under control and stable, the next phase of a QMS is to work on quality <u>continuous improvement</u>. Its purpose is to identify the opportunities to improve the existing performance of a product, process, or service to a higher standard. The common tools for continuous improvement and problem solving are reviewed in Chapter 8.

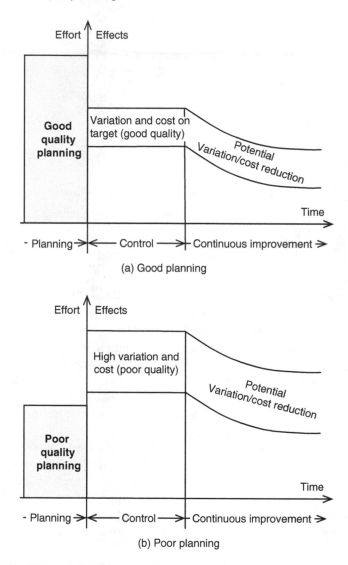

**Figure 1.8** Effects of planning on quality control and continuous improvement.

These three QMS pillars can be equally important to a quality system success. In practice, it is often quality planning, as the proactive cornerstone, that becomes the weakest link of a QMS. By design, this book focuses on this proactive pillar of QMS, bearing in mind that many approaches and tools of quality planning can apply for the other two pillars (quality control and continuous improvement).

### 1.2.2 Discussion of QMS

#### Quality Department and TQM

To dedicate and implement QMS and its processes and tasks, a quality department is an integral part of an organization. A typical setting is that quality functions are under a vice president or "Chief Quality Officer" of an organization and each product/division has a quality team or responsible person. An example of a quality department and its responsibility in a matrix type of organization is shown in Figure 1.9.

A quality department is responsible for developing and enforcement of quality policies and standards. Three fundamental questions for the quality of leadership and professionals in an organization are:

1. What is the relationship between quality and customer satisfaction?
2. What is the value added and profitability of quality efforts?
3. What are the individual employee's roles to quality in their work?

In many organizations, the answers to these three questions may be unclear. One must understand these questions and resolve them to do quality work effectively. A quality department and personnel are also responsible for the routine quality functions, e.g. inspections, process control, analysis, leading continuous improvement, etc. A survey was conducted, based on 211 Swedish quality managers, to determine what type of quality department had the best impact on an organization (Gremyr et al. 2019). The authors discovered that the most effective organization is that of one which acts as an orchestrator.

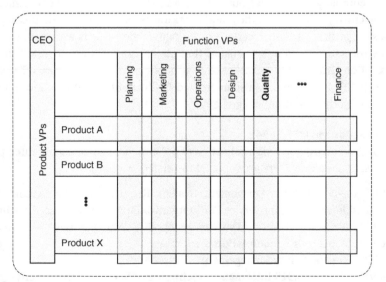

**Figure 1.9**  Quality department in a matrix organization.

Beyond the functions and efforts of a quality department, employee involvement and commitment is a core of total quality management (TQM). TQM is a participative, systematic management approach to planning and executing quality processes. Under the TQM model, quality is not only the responsibility and central occupation of a quality department, but also the role and responsibility of all employees at all levels, particularly frontline workers. Effective communication is a major enabler to motivate and involve employees in this endeavor (and will be a driving discussion in Chapter 2).

In addition to the <u>employee involvement</u>, the characteristics of TQM also include:

- <u>Customer centric</u>. Customer satisfaction, as a primary goal of a business, determines whether the quality efforts are worthwhile and successful. Driven by customer expectations, continual improvement efforts become a large aspect of TQM (Chapter 3).
- <u>Process oriented</u>. Quality is not only based on the results and outcomes but also on the process (means and techniques) to assure and improve the quality and effectiveness of work (Chapter 4).
- <u>Systematic approach</u>. Developing and implementing proactive and systematic approaches, e.g. failure modes and effects analysis (FMEA) and partnership with suppliers, can make quality work effective (Chapters 5 and 6).
- <u>Data driven</u>. Quality work is guided by reliable data (or facts) that is collected and analyzed (Chapters 7 and 8). Decisions are made with data support.

Beyond these attributes, quality management should be built on the standardization of all quality processes and tasks. A large, complex system, e.g. an automotive company, has various business operations, e.g. functional departments, manufacturing plants, facilities, and working with suppliers. All these processes, regardless of their locations or by whom the work is executed, should follow the same standards to ensure that all stakeholders are invested in the cause of ensuring quality excellence as a habit, not a series of one-off events (Zhou 2012). Furthermore, a standardized QMS should itself be continuously refined with the best practices globally.

### QM Standards and IATF 16949

The QMS standards include a collection of policies, processes, documented procedures, and records. Here are some QMS-related standards:

- IATF 16949 Quality Management System Requirements for Automotive Production and Relevant Service Parts Organizations (to be discussed further)
- ISO 13485:2016 Medical Devices – Quality Management Systems – Requirements for Regulatory Purposes
- ISO 22000 Food Safety Management
- ISO/IEC 20000-1:2018 Information Technology – Service Management – Part 1: Service Management System Requirements

- ISO/IEC 27001 Information Security Management
- AS9100D Quality Management Systems – Requirements for Aviation, Space, and Defense Organizations

A well-known industry QMS standard is ISO/TS 16949, which was created for the automotive sector in 1999. The objective of the QMS standard is to harmonize the assessment and certification schemes in the supply chains of the automotive industry worldwide. The principles and methods of the ISO/TS 16949 standard may also be a valuable reference for other industries. ISO/TS 16949:2009 is replaced with IATF 16949:2016. The latest standard shares the seven quality principles:

1. Customer focused: To ensure organizations meet customer requirements and strive to exceed customer expectations.
2. Leadership: To create and maintain the environment and culture so employees are involved in achieving the organization's quality objectives.
3. Engagement of people: To ensure employees' understanding and have the tools to contribute to the organization's success.
4. Process approach: To focus on effective transformation processes from inputs to outputs.
5. Improvement: To strive for continuous improvement.
6. Evidence-based decision making: To use the analysis of data and information for decision making.
7. Relationship management: To form interdependent and a mutually beneficial relationship between an organization and its suppliers to create value for customers.

Based on these seven principles, IATF 16949:2016 is organized into 10 chapters, listed in Table 1.4 (IATF 2016). In addition, a widely used plan–do–check–act (PDCA) process is marked to the corresponding principles of the standard in the table as a practice reference.

### 1.2.3 Quality Target Setting

**Target Setting Process**
An organization and the development projects of a product or service should have a quality goal or target, which is often defined by a qualitative statement, such as:

- "Our quality goal is simple: Making your product, to your standards, every time" (East West 2020).
- "The TI quality goal is to ensure that its products meet customer expectations" (TI n.d.).
- "Patient safety – having the right systems and staff in place to minimise the risk of harm to our patients and being open and honest and learning from mistakes if things do go wrong" (Guy 2020).

**Table 1.4** Main contents of IATF 16949:2016.

| Chapter (Section) | Title | Content | PDCA |
|---|---|---|---|
| 1–3 | Introductions | Scope, normative references, terms, and definitions. | NA |
| 4 | Context of Organization | Requirements for interested parties and their needs and expectations. Definitions of the requirements for determining the scope of a QMS and general QMS requirements. | Plan |
| 5 | Leadership | Leadership commitment to a QMS, corporate responsibility, and quality policy. | |
| 6 | Planning | Risks, opportunities, and risk analysis. Requirements for preventive actions, contingency plans, objectives, plans, etc. | |
| 7 | Support | Requirements for people, infrastructure, work environment, resources, knowledge, competence, awareness, communication, etc. | |
| 8 | Operation | Requirements on planning, product review, design, purchasing, creating the product or service, and controlling the equipment used to monitor and measure the product or service. | Do |
| 9 | Performance Evaluation | Assessment of customer satisfaction, internal audits, monitoring products and processes, and management review. | Check |
| 10 | Improvement | Requirements for problem solving, corrective actions, error-proofing, and continual improvement. | Act |

*Source*: Based on IATF, (2016). IATF 16949 Quality management system requirements for automotive production and relevant service parts organizations, International Automotive Task Force

Target setting is a planning process to establish a goal, which can be presented in terms of various aspects, e.g. revenue, profit, quality, and market share. A quality target should be included in the business goals of an organization.

A target setting process follows five steps, and these steps are defined with explicit responsibility and timing, as shown in Figure 1.10. Before starting target setting, supporting and related data must be collected. After the five-step target setting process, the subsequent step is to develop an executable plan for the targets.

1. To propose <u>preliminary quality targets</u>. Preliminary targets can be based on customer expectations and rivals' performances. For example, the expected manufacturing quality of a product is 95% without any type of repair or rework, which is also called first-time quality or first-time-through quality.
2. To do a <u>data analysis</u>, based on the current performance, resources, etc. to understand and justify the selected preliminary targets.

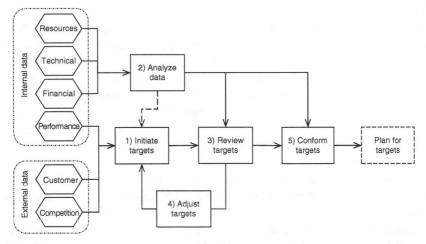

**Figure 1.10** Target setting process flow.

3. To <u>review</u> the preliminary targets, based on the analysis results of internal and external data. This step is the core of a setting process to have aggressive, yet feasible targets.
4. To <u>adjust</u> the targets if the preliminary ones are deemed too easy or aggressive to achieve for a period, e.g. in a year, if necessary. The adjusted targets must be supported by the data.
5. To <u>confirm</u> and approve the targets for prosecution.

**Considerations in Target Setting**

The considerations for a quality objective include being specific, measurable, achievable, relevant, and time-bound. Combined with objective measurability, a mnemonic acronym specific, measurable, achievable, relevant, and time-bound (SMART) may be used for all five aspects.

A key to the success of quality planning and execution is target measurability. A qualitative quality goal can be vague because of the difficulty of measuring it. For example, it is unclear how to measure "customer expectations" and "minimized" risks without some sort of quantifiable metric or dataset for analysis. Thus, it is recommended that a target statement be quantitatively defined. Here are a few examples of targets with measurable criteria:

- "Quality goal = 40 ppm (815k pcs annual volume) for all products" (Futaba 2017).
- "Achieve inpatient HCAHPS (Hospital Consumer Assessment of Healthcare Systems and Providers) overall rating and the overall quality of care score at or above the 75th percentile" (Sparrow 2019).
- "On-time Deliveries – we seek to have an on-time delivery rate of 95% or higher." (Genesis n.d.)

Target setting is predictive in nature, so the targets may not be perfectly accurate with some of the uncertainties inherent to a business environment. Often, a target is a bit aggressive aiming at the industry benchmark or best in class, as an easy one may not be very meaningful. Target setting must consider multiple constraints and feasible factors, and is subject to adjustment. For example, a national pizza chain Dominos made its delivery goal too aggressive – "30 minutes or it's free" – in the 1980s and was largely discontinued in the early 1990s (Janofsky 1993). Figure 1.11 illustrates an example of the current status, next-year target, and benchmark for target setting. In addition, a target can be set as a range, while the upper limit may be a stretch goal.

The ambitious and achievable quality goals can be set by applying the Hoshin planning approach (discussed in Chapter 2) with good communications. For a complex product or service, the system quality target needs to break down scientifically to its main elements and/or realization processes. For example, a car can only satisfy its quality targets if all of its functions are themselves of good quality. In addition, achieving a target is a process. Frequent reviews on the progress of the quality goals are necessary, particularly for aggressive goals. Quality verification and validation are discussed in depth in Chapter 4.

### 1.2.4  Cost of Quality

#### Types of Quality Cost
In addition to the technical dimensions of quality above mentioned, financial aspects related to quality are important, as they are in line with the financial and non-financial performances of an organization. Understanding the relationship between quality and cost (both up-front investments and operational

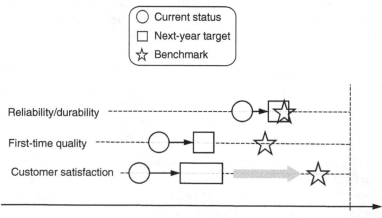

**Figure 1.11**   Targets vs. current performance levels and benchmarks.

expenditures) can encourage managers and employees to pay close attention to quality planning and operational effectiveness.

Cost of quality can be discussed in different ways. The cost comes from the investments and efforts to ensure good quality and the consequences of poor quality (see Figure 1.12). There are many elements related to these good and poor quality categories. The kind and number of elements depend on the types of a product and service.

The cost of *good* quality is the investments in early design phases and later operation phases. For example, developing a quality function deployment (QFD, discussed in Chapter 3) and failure modes and effects analysis (FMEA in Chapter 5) must be planned and conducted in development phases as up-front investments. Supplier selection, certification, and monitoring are the tasks to prevent and control supplier quality issues (Chapter 6). Maintenance is an important investment to operational quality and effectiveness. The up-front efforts and investments for good quality can be a challenge to development budget, time commitment, and may sometimes clash with traditional corporate culture.

In most cases, the cost of *poor* quality is incurred in the operation and execution of a project or service, and is directly measurable. It is called the cost of "poor" quality, because the efforts and expenditures are for the quality issues or defects already occurred. It is important to understand that most quality issues and defects can be avoided by being proactive and doing preventive tasks. For example, lack of proper maintenance has a direct relationship to some types of

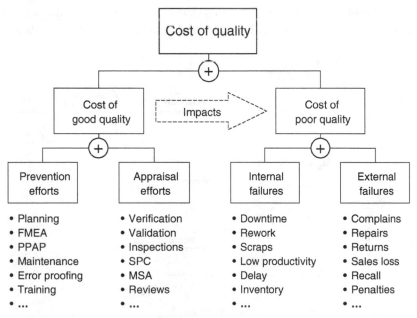

**Figure 1.12** Elements of cost of quality.

poor quality. In other words, the efforts and investments for good quality can significantly reduce the cost of poor quality. That is, an ounce of prevention is worth a pound of cure (Franklin 1735).

**Total Quality Cost**

The total quality cost can be an assessment by combining both good and poor quality costs. Figure 1.13 shows a general relationship between the total quality cost and good and poor quality costs. The figure also illustrates an overall negative correlation between the cost of good quality and the cost of poor quality.

The total quality cost has the lowest point for a specific product or service, shown as "economic quality" in the figure. For most luxury brands of products and services, the companies strive for the lowest failure rates or "best quality" (illustrated in the figure) possible even with a high total quality cost. Some business operations, such as airplane transportation and surgical procedures, must be at the highest safety quality possible, which comes at a price premium. Most business operations are in between the "economic quality" and "best quality" ranges. A key question is where the best balance point is when considering the total cost of quality and the quality to customer satisfaction for a product or service.

Quality planning affects the quality performance and quality cost. In general, the prevention and appraisal efforts are planned in the early phases of product and service development to reduce failures and defects. The quality assurance from planning efforts and approaches is sometimes called "built-in quality," and this concept is detailed in depth in later chapters.

Considering the total quality cost makes economic sense. A challenge is to quantify the total cost curve in the figure and know the approximate locations

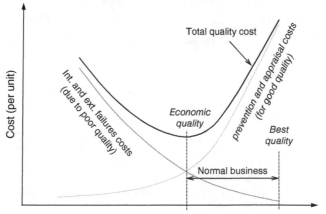

**Figure 1.13** Economic quality and best quality.

of "economic quality" and "best quality" for a product or service. Such questions require a dedicated data analysis, based on historical data and market benchmarking. Based on a total cost curve, management can make an informed decision on quality investments and quality targets. Another issue in putting this concept to practice is an up-front financial constraint. It can be difficult to allocate a sufficient budget on good quality costs to ensure that a product or service meets its quality goals in an effective way.

## 1.3   Quality Planning

### 1.3.1   Planning Process Overview

#### Definition of Quality Planning

Dr. Juran described quality planning as a systematic process for developing services and processes that ensure customer needs are met (DeFeo 2016). Juran's definition was originally for manufacturing industries, but its concept is suitable for all types of business operations when their primary purpose is to serve their customers.

Quality planning is vision and goal driven. Visions and goals are long-term and mid/short-term objectives, respectively, for the value of products and services, based on customer expectations. A vision is normally set at the top level of an organization. For example, "Mayo Clinic will provide an unparalleled experience as the most trusted partner for health care" (Mayo Clinic n.d.). A quality goal is more specific for a product and at department levels. For instance, a quality goal of a vehicle model is 140 problems per 100 vehicles (the US industry average in 2020 was 166 problems per 100 vehicles (Power 2020)).

Quality planning can start from the vision established and a gap analysis. The inputs to quality planning include project plan, assets (controllable resources), and capability, as shown in Figure 1.14 as a system overview. As an input, risk analysis and strengths–weaknesses–opportunities–threats (SWOT) analysis built a foundation to quality planning, which is discussed in depth in Chapter 2. Depending on a development project, more input information may be needed, such as existing similar parts, lessons learned, product mixes, quality specifications, supplier quotes, and feasibility studies.

The outputs of quality planning are the deliverables used to meet goals and the processes used to achieve those goals. The experience of professionals, cross-functional teamwork, and corporation culture all play assurance roles to quality planning. Common outputs from quality planning can be one or all of the following documents (DOD n.d.):

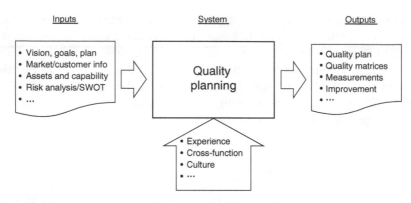

**Figure 1.14** Inputs, system, and outputs of quality planning.

1. <u>Quality management plan:</u> The plan addresses how program management will implement the organization's quality policy and achieve quality goals. Two samples of quality plan reports are shown in Figure 1.15.
2. <u>Quality metrics:</u> Quality metrics detail the definitions and descriptions of how to measure quality performance around a product or service.
3. <u>Process improvement plan:</u> The improvement plan details the purpose, process flow, analysis, and targets that are conceived to improve existing customer value.
4. Other elements, such as a quality baseline and quality assurance description, may be included in a quality plan.

### Role of Quality Planning

Quality planning is an integral part of corporate strategic management, to set long-range (5–10 years) quality goals and annual quality objectives. That means that the quality management plan, quality metrics, and continuous improvement plan are established as part of a corporate strategic plan, and revisited every year for updates. Strategic planning is further discussed in Chapter 2.

The major milestones in realization for a new product or service, viewed in terms of that product or service's lifespan, are illustrated in Figure 1.16. Quality planning efforts are integrated in early phases, such as design and process. Quality planning is for the quality assurance and control of a new product or service. The results of quality planning affect multiple phases, from design to launch of a product or service and to regular operations. Without excellent quality planning, the quality execution in development phases may be envisioned as akin to "birth defects," which can be difficult to fix in later life and adversely affect development results and effectiveness.

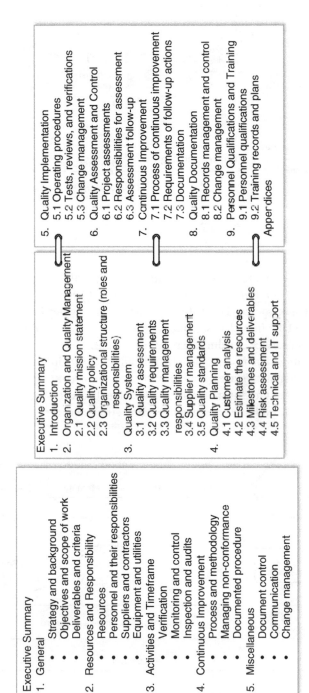

**Figure 1.15** Sample quality planning reports.

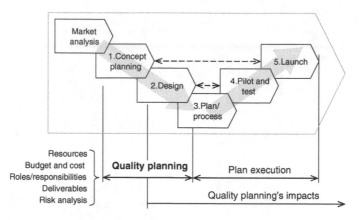

**Figure 1.16** Quality planning in product/service realization.

The functionality and role of quality planning apply to the lifecycle of a product, which is important for durable goods, such as vehicles, home appliances, and consumer electronics. The entire lifecycle of a product has five stages, shown in Figure 1.17. In such cases, quality planning has an additional segment for the after-sales service of a durable product.

Quality planning for a product lifecycle is similar, in principle and process, to that for the development stages. One important difference is that quality planning needs to account for future resource allocation, because after-sales service is related to the sales volume changing over time. After-sales services have a few

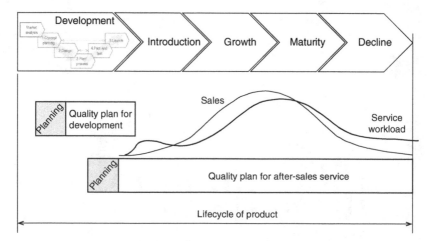

**Figure 1.17** Quality planning in lifecycle of durable goods.

major functions, such as pre-installation, user training, warranty, technical support, and return/replacement. The quality attributes of an after-sales service also include spare part availability, warranty policy, responsiveness to customer complaints, accessibility to service personnel, technical competence of service people, etc.

### 1.3.2 Considerations in Quality Planning

**Key Factors of Quality Planning**

To identify the key elements of quality planning, a study used a structured, mixed method (Tallentire et al. 2019). The authors collected data via a series of interviews and focus groups and then did inductive thematic analysis. They concluded the five key elements in a quality-planning process as shown in Figure 1.18. Note that the five elements can influence each other.

1. <u>Culture and leadership for improvement</u>. Corporate culture is linked to several dynamics, e.g. policy, teamwork, organizational structure, and communication. An effective indicator of good corporate culture is the feelings of staff, at all levels, of the trust, respect, and encouragement that they associate with their professional environment. Leadership involvement and guidance is an essential ingredient of quality planning and execution.
2. <u>Engaging and empowering staff</u>. The purpose of engaging and empowering employees is to connect their work with the organizational vision and goals. Additionally, engagement and empowerment involves the infusion of motivation, energy, and enthusiasm. For example, management encourages employ-

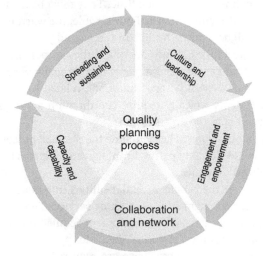

**Figure 1.18** Key factors of quality-planning process.

ees to find a link between what matters to them, and what matters to their organization. Chapter 2 will specifically discuss the Hoshin planning approach on this subject.

3. Forming collaborations and networks. In the short term, collaborations and networks are about teamwork to effectively foster and support quality planning and execution, for the sake of customers. The coordination of collaborations and networks is key for the effectiveness of planning.

4. Building improvement capacity and capability. This element is the evaluation and training of quality skills of individual employees as a building block of organizational capacity and capability. Main tasks are to assess the quality skills of employees, and help them establish individual development plans.

5. Spreading and sustaining changes. Solid achievements, e.g. improvements, new processes, revised standards, and lessons learned, are worthy references for similar situations, other departments, and even across industry. Thus, dissemination should be encouraged as contributions to a quality community at large.

**Focuses of Quality Planning**

Quality planning is an integral part of development project planning. Thus, it is vital to embed quality-planning objectives, efforts, targets, tasks, etc. into the entire development project, and coordinate with other functional teams. While planning, quality professionals address the following common items from the quality side of operations and strategy:

1. Deliverables: These include the quality goals and criteria to customer satisfaction, with a defined period or due time. They must be measurable against the predefined, satisfactory criteria. For example, the warranty cost of a new vehicle model will be reduced by 20% in the second year.

2. Roles and responsibilities: These are the duties of personnel and departments who do quality tasks and achieve quality goals. Some quality tasks and issues are complex and cross-functional, related to budget, resources, skills, etc. Such items typically require coordination and/or approval from higher management in the organization.

3. Quality assurance activities: These include the planned processes and efforts of teams to execute a quality plan within a given budget. The activities also come with defined timelines.

4. Resources: These are the people, capital, facilities, and materials required for the successful development of a product or service. The budget should be itemized and justified, considering equipment acquisition, operation, maintenance, personnel, utilities, historical data, etc.

5. Risk analysis and management: These are required for large projects, and recommended for all projects. The details of risk analysis and management are discussed in Chapter 2.

Beyond these, specific considerations may be required in quality planning for some products or services. For example, architecture and codes need to be assessed for software development, and security reviews are needed as a primary task for a network communication project.

### 1.3.3 Quality-planning Guideline (APQP)

**APQP Process**

For product development, a common approach for quality planning in the automotive industry is called APQP. APQP was developed in 1994 and revised as the second edition in 2008 (AIAG 2008). Other industries have developed their own specifications and standards, such as for software development (DOE 1997), aerospace suppliers AS9145 (SAE 2016), and environment projects (SC 2016). Some industries refer to such guidelines as a quality assurance program plan (QAPP). Here though, APQP, as a typical approach, is used as an example for further discussion.

APQP integrates quality planning into the five phases of product development (Figure 1.19). In the figure, the overall development process (Figure 1.16) is also included at the bottom of Figure 1.19. The five phases are not necessarily in purely sequential order; they can have overlaps and loops. There are various methods and tools used in the five phases, which are discussed in the following chapters.

APQP is a preventive approach to quality during development, with supporting processes of validation, evidence, and documentation. Implementing APQP may force a relatively drastic change in corporate culture and the existing practice of quality work from reactive (or firefighting) to proactive in work style and scope. For example, the quality of parts and services from suppliers are assured by the APQP approach rather than primarily by monitoring and reactions.

The APQP efforts can be organized either by part if it is unique or by part family. For a part family, limited APQP reviews and validation may be conducted to certain child parts. A product development often is on a unique part, also referred to as an engineering-to-order (ETO) product. Most service development falls into this category as well. In such cases, APQP efforts may be planned as two sections: one on common elements and features of similar parts, and another on non-standard elements and characteristics with specific reviews.

**Inputs and Outputs of APQP Phases**

A key characteristic in an APQP process is that the output items of the previous step are the inputs to the next step (Figure 1.19). The inputs and outputs of APQP phases are summarized in Table 1.5. Some outputs of a phase are just a general reference and so are the inputs to a phase, as they can be product dependent. The last column of this table lists the main sections of this book for reader quick reference.

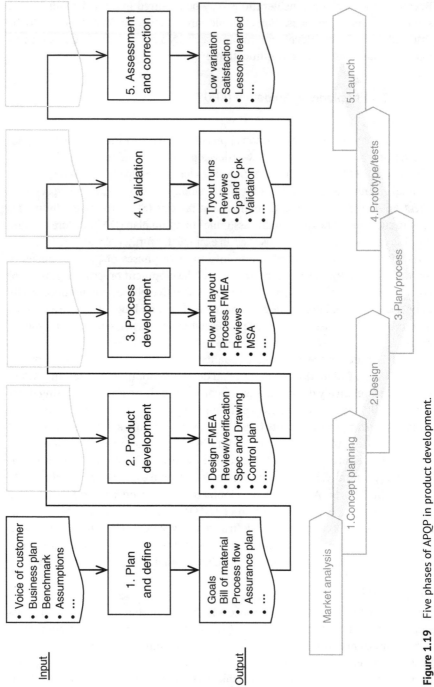

**Figure 1.19** Five phases of APQP in product development.

**Table 1.5**   Inputs and outputs of five phases of APQP.

| Phase | Function | Input | Output | This book section |
|---|---|---|---|---|
| 1 | Plan and define program | Voice of the customer<br>Business plan/ market strategy<br>Product/process benchmark data<br>Product/process assumptions<br>Product reliability studies Customer inputs | Design goals<br>Reliability and quality goals<br>Preliminary bill of material<br>Preliminary process flow chart<br>Special product/ process characteristics<br>Management support | 3.1<br><br>4.4<br><br><br>6.3<br><br>3.1<br>2.1 |
| 2 | Product design and development | | Design FMEA<br>Design for manufacturability and assembly<br>Design verification<br>Design reviews<br>Prototype build control plan<br>Engineering drawings<br>Engineering specifications<br>Material specifications<br>Drawing and specification changes<br>Equipment, tooling, facilities requirements<br>Special product and process characteristics<br>Gauge/testing equipment requirements<br>Team feasibility and management support | 5.1–5.3<br><br><br><br>4.2<br>4.1<br><br>4.2<br><br><br><br><br><br>7.3<br><br><br>3.1<br><br><br>7.1<br><br>2.1 |

Continued

**Table 1.5** Continued

| Phase | Function | Input | Output | This book section |
|---|---|---|---|---|
| 3 | Process design and development | | Shipment packaging requirements | |
| | | | Product/process quality system review | 1.2 |
| | | | Process flow chart | 6.3 |
| | | | Floor plan layout | |
| | | | Characteristics matrix | 8.2 |
| | | | Process FMEA | 5.1–5.3 |
| | | | Pre-launch control plan | 5.4 |
| | | | Process instructions | |
| | | | Measurement system analysis plan | 7.1 |
| | | | Preliminary process capability study plan | 7.2 |
| | | | Management support | |
| 4 | Product and process validation | | Significant production run | 4.2 |
| | | | Measurement system evaluation | 7.1 |
| | | | Preliminary process capability study | 7.2 |
| | | | Product part approval | 6.1–6.4 |
| | | | Production validation testing | 7.2 |
| | | | Packaging evaluation | |
| | | | Production control plan | 5.3 |
| | | | Quality planning sign-off | 7.4 |
| | | | Management support | |
| 5 | Feedback, assessment, and corrective action | | Reduced variation | 4.4 |
| | | | Improved customer satisfaction | 3.3 |
| | | | Improved delivery and service | 8.1 |
| | | | Effective use of lessons learned/best practice | |

### 1.3.4 Service Quality Planning

In service industries, they have similar quality-planning guidelines but often in different names, for example:

- Quality Assurance Project Plans (EPA 1996)
- Quality Assurance Guidelines (NIH n.d.)
- Standard Guide for Quality Planning and Field Implementation of a Water Quality Measurement Program (ASTM 2018)

Several quality-planning guidelines have been developed within individual industries and regions. For example, in Canadian healthcare, an organization called Collaborative for Excellence in Healthcare Quality (CEHQ) developed the Guide to Developing and Assessing a Quality Plan – For Healthcare Organizations. The CEHQ emphasizes that a quality plan should include the following key factors (CEHQ 2012):

1. Be aligned with the strategic plan
2. Tie in the current quality framework
3. Be a progress from previous quality plans
4. Have a clear description and be easy to follow
5. Have measurable goals
6. Define targets of quality indicators
7. Have a formal evaluation plan
8. Be feasible, based on available resources
9. Have an influence on cultural change in quality

A quality plan may be presented in the shape of a house. Its roof is the goal and there are several pillars support the goal. Discussed before, the supporting pillars are not necessarily equally important to the goal. Figure 1.20 shows two examples of strategic quality plans: one is a university's plan (GWU n.d.) on the left, the other, a hospital's plan (JHM n.d.), in the form of a house structure, based on their original statements.

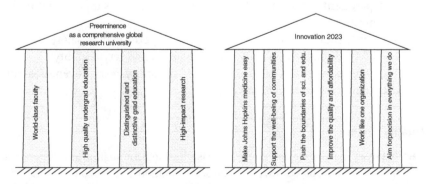

**Figure 1.20** Examples of strategic quality plan.

Juran presented quality planning and cultural transformation in five simple phases, called the Juran Roadmap (Juran n.d.), which may be viewed as a generic planning process. The five phases are:

1. <u>To decide:</u> To consider certain factors, such as competitive, regulatory, financial, and/or cultural factors, and make a decision to move forward agreed upon by the leadership.
2. <u>To prepare:</u> Senior management to prepare to articulate the strategy, direction, and plan to achieve the transformation.
3. <u>To launch:</u> To educate the organization on the transformative direction, what values to change, what support structure to assist in the transformation, and how people/performance to be measured and evaluated.
4. <u>To expand:</u> To develop the transformation into all other parts of the organization.
5. <u>To sustain:</u> To adapt and embed the new values into the business ecosystem.

As a summary, one may see that quality planning is a common practice across many professional disciplines. Quality planning practice follows a general set of principles, and has specific focuses, based on the types of business. It may be an excellent idea to review commonalities between the guidelines of different businesses, and think about improving current practices with the specific concerns in a discipline, in addition to evaluating how the different specifics for these businesses might unify, under a broader domain-wide guideline.

## Summary

### Quality Definitions

1 Quality can have a different, subjective definition for different fields and applications.
2 Quality should be customer oriented, and have multiple dimensions.
3 A department can be an internal customer if it has suppliers within an organization.
4 The dimensions of product quality include performance safety, feature, reliability, durability, aesthetics, conformance, and serviceability.
5 The dimensions of service quality include reliability, responsiveness, assurance, empathy, and tangibility.

### Quality System

6 A quality management system (QMS) is a business foundation and subsystem, built of policies, processes, procedures, information, and responsibilities.

7 The Juran trilogy QMS comprises three cornerstones: quality planning, quality control, and continuous improvement.

8 Quality targets should be specific, measurable, achievable, relevant, and time-bound (SMART).

9 Cost of quality is a vital factor for quality planning; the source categories of cost of quality include prevention, appraisal, internal failure, and external failure.

10 Total cost of quality should be systematically optimized for a product, process, or service.

## Quality Planning

11 Quality planning is vision and goal driven, with multiple inputs and outputs.

12 The outputs of quality planning include management plans, quality metrics, improvement plans, etc.

13 Quality planning starts during the early phases of development, and applies to the lifecycle of a product or service.

14 The key factors of quality planning include culture and leadership, employee engagement, teamwork, improvement capacity and capability, and sustaining changes.

15 APQP is a proactive approach. While originally designed for product development, its principle can apply to the service sectors as well.

# Exercises

## Review Questions

1 List five keywords for a definition of quality and explain why they are important.

2 Find a product or service and discuss how to measure its quality.

3 What is your definition of a QMS?

4 List the elements in ISO 9001 or a QMS standard in an industry.

5 Explain the meanings of the internal suppliers and internal customers with an example.

6 Discuss the relationship between internal customers and external customers.

7 Use an example to discuss the dimensions of a product.

8 Use an example to discuss the dimensions of a service.

9 Discuss that service quality (SQ) is $SQ = P - E$ with an example.

10 Review the role of quality planning in the Juran trilogy QMS.

11 Review the functions of a quality department in an organization.

12 Discuss a principle of IATF 16949:2016 QMS with an example.

13  Set and review a quality target for a new product or service development.
14  Explain the meanings of economic quality and best quality for a product or service.
15  List three key input factors for quality planning and discuss their roles.
16  Find the vision statement of an organization and review the quality element in that vision statement.
17  For the work of quality planning, select two focuses for discussion.
18  Discuss an appliance that has quality planning for both product development and after-sales service.
19  Explain how the outputs of an APQP phase are the inputs to the next phase.
20  Discuss if the tasks in different APQP phases can be partially overlapped.

## Mini-project Topics

1  Search sources and compare the definitions and meanings of the quality in your discipline. Please provide and justify your opinion about which definitions are more appropriate.
2  Compare the dimensions of a product and service and discuss both a common dimension and unique dimension.
3  Select a characteristic of service quality and discuss how to address it in planning.
4  One states ISO standard on a QMS is more on processes and documentation, while Juran's QMS has three functional subsystems. Review the connection between both QMS principles.
5  Review IATF 16949 for its applications.
6  Study the advantages and roadblocks of employee involvement with quality.
7  When is the best quality justifiable to a product or service when considering costs?
8  Find a quality planning report and analyze its key elements.
9  Study the link between the vision statement and quality targets of an organization.
10  Search an APQP-like procedure for a service industry.

# References

AHRQ. (n.d.). Patient safety indicators overview, US Agency For Healthcare Research And Quality, https://www.qualityindicators.ahrq.gov/Modules/psi_resources.aspx, accessed in October 2020.

AIAG. (2008). *Advanced Product Quality Planning and Control Plan*, 2nd edition. Southfield, MI: Automotive Industry Action Group.

ASQ. (n.d. a). Quality Glossary, https://asq.org/quality-resources/quality-glossary, accessed in July 2020.

ASQ. (n.d. b). What is a quality management system (QMS)?, https://asq.org/quality-resources/quality-management-system#:~:text=A%20quality%20management%20system%20(QMS)%20is%20defined%20as%20a%20formalized,achieving%20quality%20policies%20and%20objectives, accessed in July 2020.

ASTM. (2018). *Standard Guide for Quality Planning and Field Implementation of a Water Quality Measurement Program, ASTM D5612–94*. West Conshohocken, PA: ASTM International.

Benazic, D. and Varga, N. (2018). Service quality and customer satisfaction in business consulting services: An importance-performance analysis based on the partial least square method. *Proceedings of 27th International Scientific Conference on Economic and Social Development*, Rome, 1–2 March 2018. pp. 380–391.

Blessing, G. and Natter, M. (2019). Do mystery shoppers really predict customer satisfaction and sales performance? *Journal of Retailing*, Vol. 95, Iss. 3, pp. 47–62. 10.1016/j.jretai.2019.04.001

Brill, T.M., Munoz, L., and Miller, R.J. (2019). Siri, Alexa, and other digital assistants: A study of customer satisfaction with artificial intelligence applications, *Journal of Marketing Management*, Vol. 35, Iss. 15–16, pp. 1401–1436. 10.1080/0267257X.2019.1687571

CEHQ. (2012). A guide to developing and assessing a quality plan—for healthcare organizations, https://www.longwoods.com/articles/images/Guide-Developing-and-Assessing-a-Quality-Plan.pdf, accessed in April 2020.

Crosby, P. (1979). *Quality is Free: The Art of Making Quality Certain*. New York, NY: McGraw-Hill.

DeFeo, J.A. (2016). *Juran's Quality Handbook*, 7th edition. New York, NY: McGraw-Hill Education.

DOD. (n.d.). Planning for quality, https://dodcio.defense.gov/Portals/0/Documents/DODAF/Vol_1_Section_9-3-9_Quality_Planning.pdf, accessed in April 2020.

DOE. (1997). Software quality assurance plan, US Department of Energy, https://www.energy.gov/sites/prod/files/cioprod/documents/qa_plan1.pdf, accessed in June 2020.

Dutt, C.S., Hahn, G., Christodoulidou, N. et al. (2019). What's so mysterious about mystery shoppers? Understanding the qualifications and selection of mystery shoppers, *Journal of Quality Assurance in Hospitality & Tourism*, Vol. 20, Iss. 4, pp. 470–490. 10.1080/1528008X.2018.1553118.

East West. (2020). Quality, East West Manufacturing, https://www.ewmfg.com/quality, accessed in August 2020.

EC. (n.d.). Food Quality, Knowledge Centre for Food Fraud and Quality, European Commission, https://knowledge4policy.ec.europa.eu/food-fraud-quality/topic/food-quality_en, accessed in April 2021.

EPA. (1996). The volunteer monitor's guide to quality assurance project plans, US Environmental Protection Agency, https://www.epa.gov/sites/production/files/2015-06/documents/vol_qapp.pdf, accessed in April 2020.

EPA. (2019). Air Quality Index (AQI) basics, updated on June 18, 2019, US Environmental Protection Agency, https://cfpub.epa.gov/airnow/index.cfm?action=aqibasics.aqi, accessed in October 2020.

FDA. (2009). Q10 Pharmaceutical Quality System – Guidance for Industry, U.S Food and Drug Administration, https://www.fda.gov/regulatory-information/search-fda-guidance-documents/q10-pharmaceutical-quality-system, Content current as of October 17, 2019, accessed in September 2020.

Franceschini, F., Galetto, M., and Mastrogiacomo, L. (2016). ISO 9001 Certification and failure risk: Any relationship? *Total Quality Management & Business Excellence*, Vol. 29, pp. 1279–1293. 10.1080/14783363.2016.1253466.

Franklin, B. (1735). In case of fire, the Independence Hall Association, https://www.ushistory.org/franklin/philadelphia/fire.htm, accessed in March 2021.

Futaba. (2017). Meeting your needs through quality manufacturing, Futaba Corporation, http://www.futaba.com/contract_manufacturing/compliance/index.asp.html, accessed in August 2020.

Genesis. (n.d.). Quality management, Genesis Pipe & Supply, https://www.genesisoilquip.com/quality-management, accessed in August 2020.

Glodowski, M., (2019). Class Discussion, QUAL552 Quality Planning (CRN 13393), Eastern Michigan University, September 5, 2019.

Gremyr, I., Elg, M., Hellström, A. et al. (2019). The roles of quality departments and their influence on business results, *Total Quality Management & Business Excellence*, 10.1080/14783363.2019.1643713.

Guy. (2020). Guy's and St. Thomas' quality report 2019/20 – NHS, Guy's and St. Thomas' NHS Foundation Trust, https://www.nhs.uk/Services/UserControls/UploadHandlers/MediaServerHandler.ashx?id=332, accessed in August 2020.

GWU. n.d. GW's strategic plan: Path to preeminence, https://strategicplan.gwu.edu, accessed in June 2020.

IATF. (2016). *IATF 16949 Quality management system requirements for automotive production and relevant service parts organizations*, International Automotive Task Force.

ISO. (2011). *ISO/IEC 25010:2011 Systems and Software Engineering – Systems and Software Quality Requirements and Evaluation (SQuaRE) – System and Software Quality Models*, Geneva, Switzerland: International Organization for Standardization.

ISO. (2015). *ISO 9001:2015 Quality Management Systems – Requirements*, Geneva, Switzerland: International Organization for Standardization.

ISO. (2016). *ISO/IEC 25022:2016 Systems and Software Engineering – Systems and Software Quality Requirements and Evaluation (SQuaRE) – Measurement of Quality in Use*, Geneva, Switzerland: International Organization for Standardization.

Janofsky, M. (1993). Domino's ends fast-pizza pledge after big award to crash victim, The New York Times, December, 22, 1993, https://www.nytimes.com/1993/12/22/business/domino-s-ends-fast-pizza-pledge-after-big-award-to-crash-victim.html, accessed in March 2021.

JHM. (n.d.) Johns Hopkins Medicine Strategic Plan, https://www. hopkinsmedicine.org/strategic-plan, accessed in June 2020.

Juran, J. (n.d.). The five phases of the Juran roadmap, https://www.juran.com/ approach/the-juran-roadmap, accessed May 2020.

Juran, J.M. (1986). The quality trilogy – A universal approach to managing for quality. *ASQC 40th Annual Quality Congress in Anaheim*, California, May20, 1986.

Lewis, R.C. and Booms, B.H. (1983). The marketing aspects of service quality, in Berry, L., Shostack, G., and Upah, G. (Eds), *Emerging Perspective on Service Marketing*, pp. 99–107, American Marketing Association Proceedings Series, Chicago, IL.

Mayo Clinic Health System, Mission, Vision, Values. https://www.mayoclinichealth system.org/locations/la-crosse/about-us/mission, accessed in April 2020.

Minghetti, V. and Celotto, E. (2013). Measuring quality of information services, *Journal of Travel Research*, Vol. 53, Iss. 5, pp. 565–580. 10.1177/0047287513506293

Mori, R. (2018). Class Discussion, QUAL552 Quality Planning (CRN 14077), Eastern Michigan University, September 5, 2018.

NIH. Quality assurance guidelines, https://www.ninds.nih.gov/Funding/Apply-Funding/Application-Support-Library/Quality-Assurance-Guidelines, accessed in April 2020.

Nurcahyo, R. and Zulfadlillah, M.H. (2021). Relationship between ISO 9001:2015 and operational and business performance of manufacturing industries in a developing country (Indonesia), *Heliyon*, Vol. 7, Iss. 1, 10.1016/j.heliyon.2020.e05537

Parasuraman, A., Zeithaml, V., and Leonard, L. (1985). A conceptual model of service quality and its implications for future research, *Journal of Marketing*, Vol. 49, Iss. 4, pp. 41–50. 10.2307/1251430.

Power, J.D. (2020). New-vehicle quality mainly dependent on trouble-free technology, Press Release, J.D. Power, 24 June 2020, https://www.jdpower.com/ business/press-releases/2020-initial-quality-study-iqs, accessed in February 2021.

Saad, M., Abukhalifeh, A.N. Slamat, et al. (2020). Assessing the use of linear regression analysis in examining service quality and customer satisfaction relationship in premium casual restaurants (PCR) in Subang Jaya (Klang Valley) Malaysia, *Review of Integrative Business and Economics Research*, Vol. 9, pp. 369–379

SAE. (2016). Aerospace series – requirements for advanced product quality planning and production part approval process AS9145. https://www.sae.org/ standards/content/as9145, accessed in July 2020.

SC. (2016). *Quality Assurance Program Plan For The Underground Storage Tank Management Divisiona*, Columbia, South Carolin: South Carolina Department of Health and Environmental Control. https://scdhec.gov/sites/default/files/docs/ Environment/docs/DHEC%20UST%20QAPP_Rev-3.1(2).pdf. accessed in June 2020.

Sparrow. (2019). Quality initiatives, Sparrow Eaton Hospital, https:// sparroweatonhospital.org/patients-visitors/patients/quality-initiatives, accessed in August 2020.

Stevens, K. (2018). Class Discussion, QUAL552 Quality Planning (CRN 14077), Eastern Michigan University, September 6, 2018.

Sun, X., Wen, D., Yan, D. et al. (2019). Developing and validating a model of ISO 9001 effectiveness gap: Empirical evidence from China, *Total Quality Management & Business Excellence*, Vol. 30, Iss. Sup. 1, pp. S274–90. 10.1080/14783363.2019.1665867.

Syahrial, E., Suzuki, H., and Schvaneveldt, S.J. (2019). The impact of serviceability-oriented dimensions on after-sales service cost and customer satisfaction, *Total Quality Management & Business Excellence*, Vol. 30, Iss. 11–12, pp. 1257–1281. 10.1080/14783363.2017.1365595.

Tallentire, V.R., Harley, C.A., and Watson, S. (2019). Quality planning for impactful improvement: A mixed methods review, *BMJ Open Quality*, Vol. 8, e000724, 10.1136/bmjoq-2019-000724.

Tang, H., (2017). *Automotive vehicle assembly processes and operations management*. Warrendale, PA: SAE International.

TI. (n.d.). Texas instruments supplier performance expectations, Texas Instruments, https://wpl.ext.ti.com/Content/File/22, accessed in August 2020.

Toure, K., (2019). Class Discussion, QUAL552 Quality Planning (CRN 13393), Eastern Michigan University, September 6, 2019.

Tuncer, I., Unusan, C., and Cobanoglu, C. (2020). Service quality, perceived value and customer satisfaction on behavioral intention in restaurants: An integrated structural model, *Journal of Quality Assurance in Hospitality & Tourism*, Vol. 22, Iss. 1, pp. 1–31. 10.1080/1528008X.2020.1802390.

VDA, (2016). QM – System audit – Serial production, German Association of the Automotive Industry (VDA), Behrenstraße 35, 10117 Berlin.

Vuk, T. (2012). Quality indicators: A tool for quality monitoring and improvement, *ISBT Science Series*, Vol. 7, pp. 24–28. 10.1111/j.1751-2824.2012.01584.x.

Zeng, J., Phan, C.A., and Matsui, Y. (2013). Supply chain quality management practices and performance: An empirical study, *Operations Management Research*, Vol. 6, Iss. 1–2, pp. 19–31. 10.1007/s12063-012-0074-x.

Zhou, J. (2012). Excellence is not an event; it's a habit, *Quality Digest*, December 4, 2012. https://www.qualitydigest.com/inside/quality-insider-column/excellence-not-event-it-s-habit.html, accessed in September 2020.

# 2

# Strategy Development for Quality

## 2.1 Strategic Management

### 2.1.1 Overview of Strategic Management

**Business Strategy**

The short-term and long-term operations of an organization are guided by vision, mission, and goals. The vision of an organization inspirationally states the desired direction of the future, while an organization's mission is an action based statement of an organization's purpose. The goals are the general outcomes that an organization requires to achieve the vision.

Strategic development and management are broad subjects. They can be viewed as the processes, documentation, and administration involved in setting and achieving the goals and objectives of an organization. Strategic development is also called policy development or strategic planning.

In strategic development, there are several key factors, including relationships among leadership, planning processes, and workforce planning. The National Institute of Standards and Technology (NIST) administers the Malcolm Baldrige National Quality Award. Two of the Award criteria are "the relationship between the leadership system and the strategic planning process" and "the connection between workforce planning and strategic planning" (NIST 2015). The Award addresses achievements and improvements in leadership, strategic planning, customer and market focus, information analysis, human resource focus, process management, and business performance.

For a product or service, two basic strategies are differentiation strategy and cost-leadership strategy. The former is an organization meant to create better value with its products or services for customers than competitors do; the latter is an organization meant to create the same value for customers but at a lower cost. Clearly, premium quality can be a differentiation strategy for a product,

*Quality Planning and Assurance: Principles, Approaches, and Methods for Product and Service Development*, First Edition. Herman Tang.

service, or brand name, while quality is also a basic factor to a cost-leadership strategy. Therefore, quality plays a vital role in the various types of strategic development.

One contemporary method is called a balanced scorecard, originally proposed in the early 1990s. A balanced scorecard sets and measures the goals of an organization (Kaplan and Norton 1992). A typical balanced scorecard includes four connected perspectives: customer, internal processes, financial, and innovation and learning. Quality, as a part of an organization's vision and strategy, should be a driving force and indicator for all four perspectives (Figure 2.1). However, many organizations struggle with quantifying the relationship between quality and these four categories, and with quantifying the different weights of the four categories.

**Strategy Development**

There are many tasks in strategic development. The key processes include two main phases:

1.  <u>Study</u>: The study phase includes tasks of a benchmark analysis, voice-of-the-customer analysis, internal structure review, and status assessment.

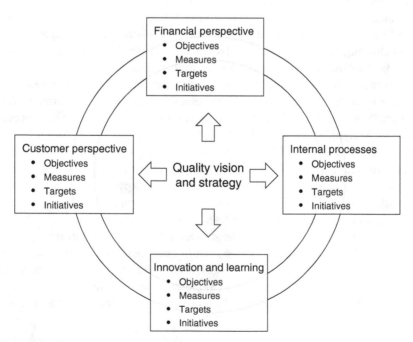

**Figure 2.1**   Quality-focusedbalanced scorecard.

2. <u>Plan</u>: The planning phase focuses on a goal setting, policy development, and implementation guideline development.

One important tool of a strategic study is the business environment analysis (BEA) (Russ 2013). In a BEA, one first identifies the factors or environment components influencing an organization's performance. In general, the factors include economical, technological, social, political, and legal components, while they are case dependent for a product or service. The factors can then be sorted as either internal or external factors, as shown in Figure 2.2. The evaluations of these factors normally come from detailed studies, such as a market research study, benchmark analysis, and/or government regulation review.

Once sorted, one analyzes the impact of these factors, and categorizes them into four groups. Such an analysis is called a strengths–weaknesses–opportunities–threats (SWOT) analysis.

- The analysis of <u>strengths</u> and <u>weaknesses</u> for an organization is primarily based on the identified internal factors, in relation to competitors. A strength is a competitive advantage in either low-cost or differentiation, or possibly their combination. Quality and innovation are part of the differentiation.
- The assessment of <u>opportunities</u> and <u>threats/risks</u> to the development and progress of an organization is mainly based on external factor analysis. Because they are external, one cannot change them, yet one may take advantage of opportunities, protect against threats, and/or reduce their impacts.

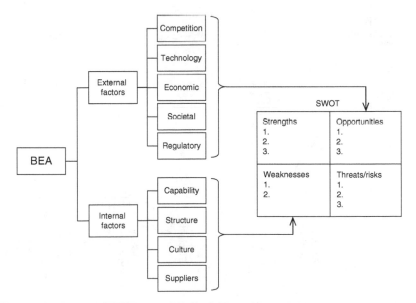

**Figure 2.2** BEA and SWOT in strategy development.

The value of a SWOT analysis is that one may clearly understand the market and an organization itself. According to a SWOT analysis, one can also plan something new to advance on products and services. Note that risk-taking is important for the development of a business, product, or service. Further discussion on risk management follows in Section 2.2.

### Employee Engagement in Strategy Development

The vision, mission, and goals of an organization are important only if they are acknowledged and pursued by all employees. As a key to the success of quality management, employee engagement can make employees feel passionate about their jobs and valuable to an organization, and motivate them to do their best work. Most organizations have a clear vision, mission, and goals in place; some organizations, large ones especially, may not marginally connect with frontline workers in terms of strategic development and implementation. As a result, frontline workers feel distanced from the planning process and have limited interests in the success of an organization's mission and goals.

A comprehensive survey study was conducted by *Harvard Business Review* (HBR 2013). The survey received 568 responses from the executives of global organizations with 500 or more employees. The survey showed that over 70% of those polled considered communication and employee engagement as the second and third important factors to business success, as shown in Figure 2.3.

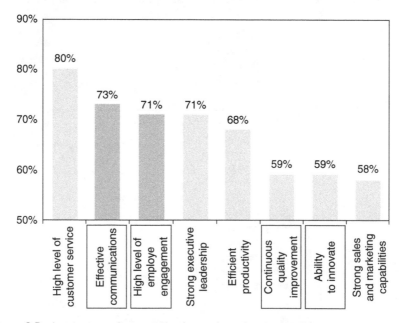

**Figure 2.3** Importance of communication and employee engagement.

One approach that helps engage employees and bring an organization together during strategy development and execution is called Hoshin planning. Hoshin is short for "Hoshin Kanri," a Japanese word meaning "strategic or policy management." In October 1988, US quality professionals visited six companies in Japan, and asked each of them about quality management tools and systems. Hoshin planning was the only system that was used by every company interviewed (King 1989).

Hoshin planning is a set of techniques to engage every person in an organization in planned objectives and continuous improvement. Everyone should be engaged in organizational planning, and the subsequent assessment of how well the plans are being realized. In Hoshin planning, the communication flows through the hierarchical structure of an organization as a dialog, as shown in Figure 2.4. Therefore, the two keywords of Hoshin planning are communication and alignment.

Hoshin planning is a process developed to ensure that vision, mission, and goals are communicated throughout an organization. To connect organization goals and employee efforts, some questions may be asked, e.g.:

"Is what I am doing aligned with my organization's goals?"

"Are the means I am using going to help achieve our organization's target?"

Hoshin planning can also be viewed as:

- A process used to address critical business needs, align organization resources, and develop the capability of employees at all levels.
- A work plan developed for every employee from top management to shop floor level. The execution of the plan is monitored.

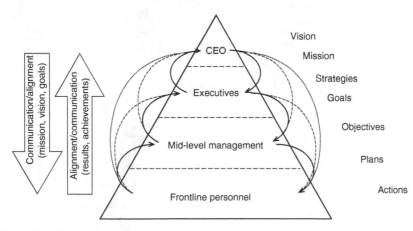

**Figure 2.4** Employee engagement in strategic plan management.

Hoshin strategic planning is a general approach and can be used for almost all types of business. For example, a Hoshin approach can be used in the first phase of advanced product quality planning (APQP) discussed in Chapter 1. Compared with other quality management processes and lean manufacturing methods, a survey study found that Hoshin planning was less used in Western academics and practices (Nicholas 2016). This implied opportunity warrants an in-depth study of Hoshin strategic planning and its implementation.

### 2.1.2 Hoshin Planning Management

**Hoshin Process Steps**

The conceptual Hoshin planning approach suggests a process comprising seven steps (see Figure 2.5). The process begins with a strategic vision at the highest level and ends with improvement objectives, actions, and reviews at the shop floor level. There are multiple communication loops among the levels.

1. <u>Establish organizational vision and mission statements</u>. The major tasks in this stage include evaluating existing policies, setting long-range (5–10-year) goals, reviewing management structure, and solidifying vision and mission statements. For example, Ford Motor Company states, "We are once again working to revolutionize mobility, fueled by new challenges and creating

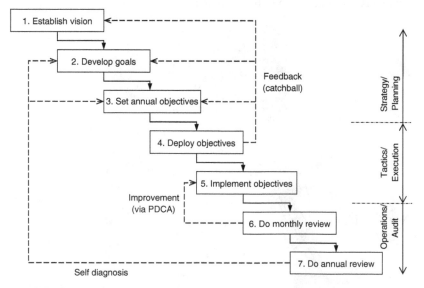

**Figure 2.5** Seven steps of Hoshin planning.

solutions to build a better world for everyone. Our mission is to drive human progress through the freedom of movement." (Ford n.d.).

2. Develop medium-term breakthrough goals. The goals can be for three to five years as a mid-range strategic plan, such as a new product line, market share expansion, and technology advance. The work in this stage may consist of four elements:
   - Objectives (targets)
   - Owners (accountability)
   - Measurable targets (measurements)
   - Major projects (means to achieve)

3. Develop annual objectives. This step is the transition point from strategy development into strategy deployment. Annual objectives should be measurable and derived from the strategic plan developed in Step 2. Additional discussion follows in the next subsection.

4. Deploy annual objectives. This is to set workable objectives at a departmental level. This step is a cascading process that brings all layers of an organization into alignment with the organizational long-term and medium-term objectives.

5. Implement annual objectives. This is a step to carry out actions that meet departmental objectives, by individual employees. Often, a maximum of five goals is set for a year to ensure all employees know what they should strive for. This limitation can prevent the "flavor-of-the-week" syndrome that tends to exhaust employees and reduce their morale.

6. Conduct monthly reviews. The purpose of a monthly review is to assess performance against predetermined objectives. The review can help diagnose issues, roadblocks, and manage the means to improve. The plan–do–check–act (PDCA) process, discussed in Chapter 8, is often used to improve performance after a monthly review.

7. Conduct annual reviews. These are individual comprehensive appraisals of the organizational and department annual objectives. Employees are self-evaluated, and appraised by superiors, against the annual objectives and core competencies that are identified. The review results of individuals and departments are a basis for the objective and plan development for the following year.

**Targets and X-Matrix Development**

In Steps 2 through 5 of a Hoshin process, a Hoshin X-matrix is used to link the steps. The four-quadrant matrix summarizes Hoshin planning on a single page, and makes the process and dependencies transparent. A generic example of a Hoshin X-matrix is shown in Figure 2.6. Hoshin X-matrices may be different for various applications, such as healthcare (Barnabè and Giorgino 2017).

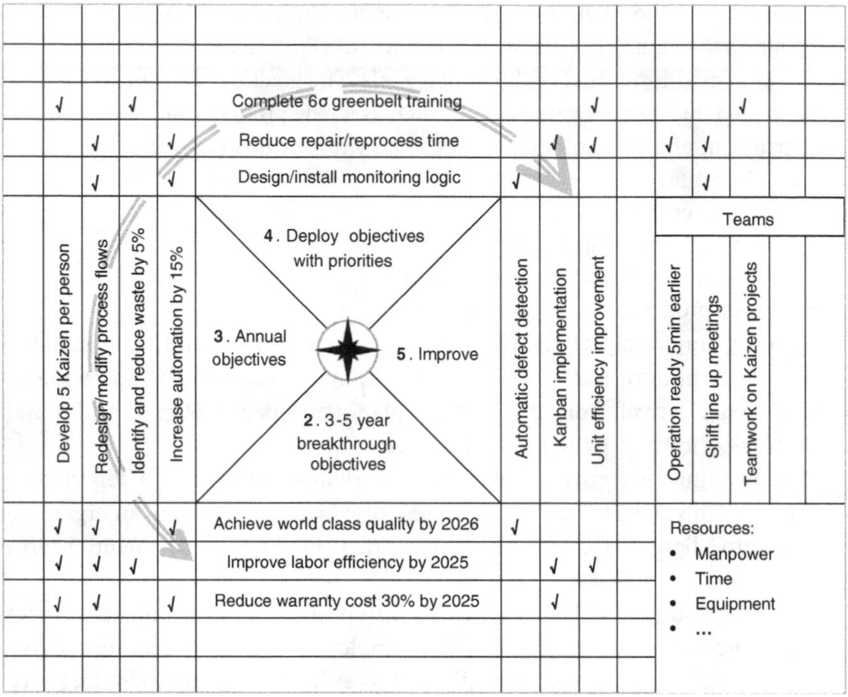

Center diagram (Hoshin X-matrix quadrants):
- 4. Deploy objectives with priorities
- 3. Annual objectives
- 5. Improve
- 2. 3-5 year breakthrough objectives

Top right: **Teams**

Bottom right:
Resources:
- Manpower
- Time
- Equipment
- ...

| | Develop 5 Kaizen per person | Redesign/modify process flows | Identify and reduce waste by 5% | Increase automation by 15% | Automatic defect detection | Kanban implementation | Unit efficiency improvement | Operation ready 5min earlier | Shift line up meetings | Teamwork on Kaizen projects |
|---|---|---|---|---|---|---|---|---|---|---|
| Complete 6σ greenbelt training | √ | | √ | | | | √ | | √ | |
| Reduce repair/reprocess time | | √ | | √ | √ | √ | | √ | √ | |
| Design/install monitoring logic | | √ | | √ | √ | | | | √ | |
| Achieve world class quality by 2026 | √ | √ | √ | | √ | | | | | |
| Improve labor efficiency by 2025 | √ | √ | √ | | | √ | √ | | | |
| Reduce warranty cost 30% by 2025 | √ | √ | | √ | | √ | | | | |

**Figure 2.6**  Target and means (Hoshin X-matrix).

After the organizational vision (Step 1) is completed, the following four process steps in Hoshin planning have the corresponding quadrants:

Step 2 – Develop medium-term breakthrough objectives (south).
Step 3 – Set annual objectives based on the breakthrough objectives (west).
Step 4 – Deploy objectives with top priorities (north).
Step 5 – Implement to improve on the core metrics (east).

Step 3 is the core of Hoshin planning, where one sets annual objectives for each department and individual. The annual objects are the foundation used to reach the three–five-year breakthrough objectives. This is also the point where a Hoshin X-matrix is initiated, to begin strategy and objective deployment.

At the departmental and individual levels, Step 4 involves multiple tasks for deployment and implementation to concentrate on for the current year. These tasks include establishing specific targets, and identifying necessary means. In

Steps 3 and 4, it is vital that all employees are aligned with the organizational vision and goals, and that their performances are measurable.

In the Hoshin X-matrix example in Figure 2.6, the interaction and interdependent relationship between the tasks in different steps (or quadrants) are shown using the check mark "√." At the department or individual levels, the relationship may be quantified by using a percentage, a weight in a scale of 1–5, or a score (Akarsu et al. 2018).

There are other considerations to Hoshin planning and X-matrix development:

- The development process follows the planning steps, i.e. clockwise starting from the south quadrant, while the catchball feedback process (discussed in the next subsection) is in a counterclockwise direction.
- On the far-right side of an X-matrix, the people (owners) responsible for executing the plan can be identified. Sometimes, the related resources are listed there as well.
- Hoshin planning also addresses policy, process, and teamwork and focuses on root causes when issues occur and operations not reaching the predefined objectives based on reviews.
- Hoshin planning may not be difficult to carry out, yet the resultant X-matrix can be impressive. As a tool, Hoshin planning can have a high usefulness potential if it becomes part of normal business strategy reviews and drives meaningful actions.

### Catchball – Engagement Process

Conventional management focuses on declaring objectives and checking results. Traditionally, a manager directs their subordinates what to do, and when to finish. However, most people dislike being told how and what to do and may have their own ways to do tasks and reach their goals. Thus, the traditional management style may not be the optimal way to motivate employees to do their best.

Industrial practices and academic research show that employee's satisfaction is a prerequisite to customer satisfaction. A study based on the data of 2,220 Portuguese small and midsize companies showed the correlation between employee satisfaction and customer satisfaction was 0.69 (Fonseca and Ferro 2016). A survey from 261 customers and 261 managerial employees showed that focusing on employee engagement could lead to a favorable corporate image and enhance customer satisfaction (Zameer et al. 2018). Another study concluded, "From a macro perspective, employee satisfaction trajectories strongly affect customer satisfaction for companies with significant employee – customer interaction but not for companies without such interaction." (Wolter et al. 2019).

In Hoshin planning, there is a vertical loop communication process, called "catchball" in Steps 1 through 4. The process suggests that the higher levels of management set a goal and maintain an open dialog for feedback. The management "tosses" the goals, like a ball, to the lower levels and waits to receive feedback. Subordinates catch this ball and then throw it back to management, with executable plans and concerns.

The overarching goal of the catchball process is to have a good alignment throughout an entire organization. At the different levels, the engagement catchball dialog has different focuses, as already discussed (Figure 2.4). Sometimes, a catchball process may require several attempts to work correctly, though after adjustments, this system will succeed in its intended purpose.

The catchball process can boost the acknowledgment and motivation of lower-level personnel and their commitment to implementing organizational plans to achieve the goals. The process is an effective way to get everyone involved in planning and enable a consensus among teams in terms of objective settings and measures. In addition, the process encourages everyone to think about their individual long-term career goals.

### 2.1.3 Implementation Considerations

**Factors for Implementation**

There are various factors to be considered in Hoshin applications. Based on 101 published studies between 1991 and 2015, Silveira et al. (2017) summarized 20 guidelines, in six categories, for Hoshin planning implementations. These factors and guidelines can be a reference when implementing a Hoshin process (Figure 2.7).

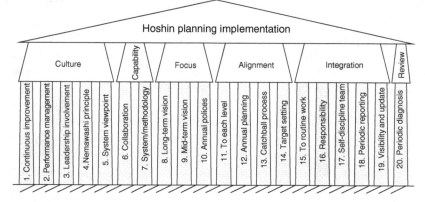

**Figure 2.7** Guidelines for Hoshin planning implementation.

<u>Organizational culture:</u> Culture items include continuous improvement mind-set, leadership involvement, and systemic viewpoints.

1. To establish a culture of continuous improvement combined with a culture of learning organization.
2. To develop management for results with challenging targets to drive high-impact performance improvements.
3. To actively involve leadership at all levels of the organization to foster mobilization, alignment, and commitment, and to ensure the effective use of management methodologies.
4. To use the Nemawashi principle (informal communication to lay the groundwork for a proposal) to deal with the conflicts in decision-making.
5. To develop a systemic view to reach its results without harming the results of other areas and the overall results.

<u>Capabilities:</u> These include problem-solving skills, tools, resources, and structures.

6. To develop a collaborators' orientation to troubleshooting and teamwork.
7. To develop the ability to address systemic issues by using cross-functional teams and structured methodology.

<u>Focus:</u> Centered on long-term vision, medium-term goals, and annual objectives.

8. To have a long-term vision meaningful to the employees at all levels.
9. To translate a long-term vision into medium-term goals.
10. To translate medium-term goals into vital annual policies targeting both the business growth and improvement of its critical areas.

<u>Alignment:</u> The catchball process implementation is a key for an alignment process.

11. To design the annual policies at each level to be meaningful to work teams, balanced in relation to performance dimensions, and sufficient to achieve the overall results.
12. To involve everyone to a greater or lesser degree in an annual planning.
13. To deploy each policy using a catchball process between the different levels to foster the mutual adjustment of strategy planning.
14. To define targets and means based on cause-and-effect analysis.

<u>Integration:</u> To ensure targets are integrated into routines, cascaded deployment, self-discipline, periodic reports, and visibility.

15. To integrate targets into the routines of process management or project management.
16. To assign cascaded deployment the responsibility for the management and achievement of the targets.

17. To develop self-discipline in work teams for the completion of the activities within daily routines.
18. To develop periodic reports on performance.
19. To keep the key information of strategy planning and execution visible and updated.

<u>Review:</u> Describes a periodic review, and subsequent diagnosis, by senior management that is performed to understand the roadblocks or challenges in achieving targets.

20. To conduct a periodic diagnosis by senior management to hear and understand the teams' difficulties in achieving targets and provide support.

**Teamwork**

The core value of Hoshin planning is teamwork across all levels – vertical coordination for true communication, alignment, work assignments, and goal setting. Please note that vertical coordination work for strategic planning requires additional efforts and time. Tennant and Roberts (2001) showed that a catchball process justified the initial investment for the development at a company over 10 years.

The success of bottom-up, vertical communication requires two key components. One is that employees understand their roles in an organization's strategy and are motivated to participate in strategy development. The other is the conversation mechanism, i.e. employees have channels through which to communicate. In addition to regular face-to-face meetings, other types of communication methods include an anonymous survey, online suggestions box, focus group discussion, frontline forum (a senior leader gathers together a representative group of frontline staff), etc. Various methods help with bottom-up communication, of the voice of employees from lower levels in an organization.

Implementing a Hoshin process and developing an X-matrix should be done as a team. For each business unit, the unit head should take the responsibility to guide and maintain the departmental Hoshin X-matrix. An annual review is needed to check measured performance against the predetermined objectives, and revise the X-matrix as specified in Steps 6 and 7 of Hoshin planning, when necessary.

Implementing Hoshin planning should be an annual event to update critical issues, communicate, and align frontline employees with the company goals via a catchball process. For example, when summarizing a missing target, one should brainstorm the reasons a plan worked or did not work, and what can be done differently in the next year. Here are a few more examples of Hoshin applications:

- <u>Healthcare</u>: Improving customer satisfaction of a healthcare facility – reading the customers' needs (Gonzalez 2019).
- <u>Quality management</u>: Evolution and future of total quality management – management control and organizational learning (Dahlgaard 2019).

- **Manufacturing**: Cellular and organizational team formations for effective lean transformations (Vries and Van der Poll 2018).
- **Education**: Hoshin X-matrix drives engineering leadership program success (DeRuntz et al. 2014).

### Possible Challenges

Strategic planning has been implemented in many organizations, but not every plan is effective. Common issues with strategic planning include:

1. Ambitious and unrealistic goals
2. Inaccurate assumptions
3. Difficult to measure
4. Not comprehensive (or fragmented)
5. Weak link between long-range goals and routine execution
6. Fragile – having to change frequently

For an organization, even one issue can make planning ineffective or even lead to plan failure. Thus, it might be a good idea to develop a checklist based on these potential issues, to reassess current practices prior to revision of an existing strategic plan.

When doing strategic planning, common challenges include:

1. Additional workload (meetings, paperwork, etc.) for middle and first-line management.
2. Difficulty of information (particularly undesired or disagreeable information) moving up in a command chain.
3. Reactive (non-proactive) mentality and work style (or corporate culture) in middle and frontline management.
4. Resource constraints, e.g. staff and technology availability, to pursuing resolutions.

It can be difficult to set up aggressive yet realistic goals. For example, if corporate management wants their products to rank top for quality in the industry worldwide in five years, then that management needs to have a deep understanding about the "shop floor," roadblocks, and capabilities to know how feasible this goal is. For instance, materials supply and product distribution chains might be excellent, but is the equipment in the workshop equipped to build the improved product? Are frontline workforce and management fully trained for problem-solving on the new standards?

Incorporating company value ideologies into frontline employees may take a considerable amount of time and effort. It is crucial to show them over time that contributing toward organizational goals benefits their interests, such as job security, career progression, and financial motivators. It can take a month, at a department level, to communicate with all employees, prepare feedback, and develop a plan for goals.

Transferring multiple-year goals into annual objectives (Step 3 in the Hoshin process) and then into departmental or operational objectives (Step 4) requires more work than just a waterfall event in a hierarchical organization chart. There is a lot of cross-functional, cross-departmental coordination, and interaction in order to develop a plan to reach a goal, e.g. a new product or a profit target. For medium-sized and larger companies, the standardization of a Hoshin process is very important. Based on the seven process steps and X-matrix suggested earlier, an organization may adapt and develop its own process steps and forms.

Resolving challenges and issues may be much more difficult than identifying them; there are no universal solutions. One would need to work diligently to address each unique challenge to remedy a given situation.

## 2.2 Risk Management and Analysis

### 2.2.1 Risk Management Overview

#### Principle of Risk Management

Risks may be thought of in a negative sense as events that could have some adverse impact on a product, service, or project should that risk occur. Risk management is a general business approach and process. Its purposes are to identify, evaluate, and create a plan to handle potential problems and to achieve business goals. The efforts of risk management also include coordinating resources to mitigate the impacts of risks, monitor identified risks, and control risk probability. Therefore, risk management is important to an organization, particularly those of innovation, or those aggressively pursuing a new product or service.

ISO 31000 is a standard family relating to risk management (ISO 2018). This standard family includes ISO 31000:2018 Principles and Guidelines on Implementation, ISO/IEC 31010:2019 Risk Management – Risk Assessment Techniques, and ISO Guide 73:2009 Risk Management – Vocabulary. Based on the ISO standards, risk management includes 11 principles (ISO 2007), as illustrated in Figure 2.8. Among them, the fourth principle, "Explicitly addresses uncertainty" is a core of risk management, while the other principles are shared with other functions of business management. During the development and execution of risk management, one may apply all or some of the principles when tackling a given specific case.

Uncertainty is about imperfect or unknown information in some environments, or associated ambiguity with a given identity. For quality planning in

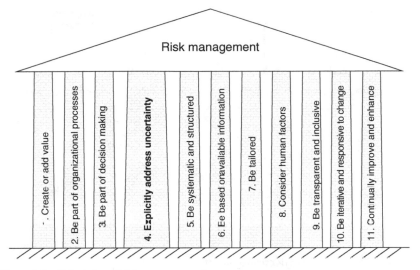

**Figure 2.8** Principles of risk management.

terms of a new product or service, uncertainties include a market trend, variation of customer expectation, adoption of new technologies, reliability of data collection, etc. These uncertainties should be first identified and then addressed in risk management, prior to new product or service development.

Risk management and analysis are widely conducted in various industries. Here are a few examples:

- Engineering design: Principles of risk-based rock engineering design (Spross et al. 2020).
- Services: Developing a hospital disaster risk management evaluation model (Arab et al. 2019).
- Government: Social risk, fiscal risk, and the portfolio of government programs (Hanson et al. 2019).
- Business: A proposed operational risk management framework for small and medium enterprises (Naude 2017).

**Process of Risk Management**

The risk management process normally has six steps, as shown in Figure 2.9 and briefly explained in Table 2.1. The process of risk management overall is similar to PDCA, while the plan also includes additional work on the risk identification (Step 1) and analysis (Step 2). Sometimes, the steps of a risk management process may be reduced to five or even four by combining some tasks, such as the execu-

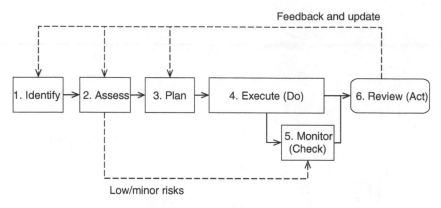

**Figure 2.9** Overall process flow of risk management.

**Table 2.1** Process steps of risk management.

| No. | Step | Task and Explanation | PDCA |
|---|---|---|---|
| 1 | Identification (what) | To identify and define potential threats and risks that may negatively influence a specific objective, process, and/or performance. Some risks can be identified in a later phase. | |
| 2 | Assessment (how much and why) | To identify the link between the risks and targets, estimate the likelihood of a risk occurring, understand the severity of its potential impacts and consequences. Often in multiple facets, e.g. timing, quality, cost, etc. | Plan |
| 3 | Examination and planning | To analyze and rank the risks and develop countermeasures and a work plan for the high-ranked risks, including mitigation processes, prevention tactics, contingency solutions, etc. For low risks, to keep monitoring them. | |
| 4 | Execution and/ or treatment | Just to execute the plan of risk countermeasures and treatments developed from Step 3. | Do |
| 5 | Monitoring | To continuously track the identified risks, countermeasure tasks, and be sensitive to new risks. For minor risks, to monitor their effects and validate the decisions of no treatment. | Check |
| 6 | Review and update | To review the risk treatments and monitoring output, and update the risk management performance periodically. | Act |

tion and monitor steps. Regardless of the number of steps, a risk management process retains the same flow to identify, analyze, and contain risks.

### 2.2.2 Risks and Treatments

**Types of Risks**

In many cases, one needs to distinguish between threats and risks. A threat is an expression that likely causes damage to an organization or product, while a risk is the possibility that a harm will be realized. Threats can stem from a variety of sources, including financial uncertainties, legal liabilities, strategic management errors, accidents, and natural disasters. A threat can be converted to a risk by understanding the threat as a vulnerability of a product or service. Threats are normally external and may not be under an organization's control. However, one needs to find a means to reduce and manage the associated risks of these external threats.

Given the close relationship between threats and risks, they may be considered together in risk management. Threats and associated risks may be categorized into four groups, as shown in Figure 2.10. While there are many items, factors, and types of risks, some of them overlap, such as predictable and technical changes.

For individual risks, one needs to understand their mechanism (how it causes harm), consequences (what type), severity (how bad), and likelihood (probability). There are various ways to study risks. One method is failure mode and effects analysis (FMEA), which is discussed in Chapter 5. Other tools, e.g. the

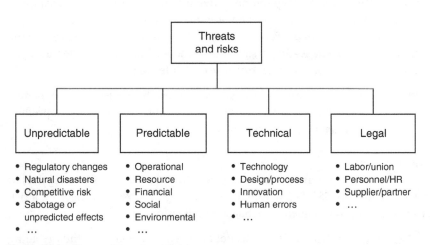

**Figure 2.10** Types of threats and risks.

fishbone diagram and 5-Why approach (discussed in Chapter 8) are often used. Special attention may be given to the following items to address risks:

- New designs and significant modifications, particularly for those with technical challenges.
- Features and characteristics critical for meeting customer expectations, such as warranty and major quality issues in current operations.
- External suppliers and incoming materials with potential impacts on the operations and quality.
- Unplanned changes during project execution, such as a timing reduction, or cost cut.

Another major risk, called product or service liability, can cause consumer injury by the manufacturer's product or provider's service. The liability is the legal obligation of a manufacturer and service provider to compensate for an injury or damage caused by a defective product or service. In early development phases and quality planning, potential high-impact quality issues, particularly those concerning safety, should be identified, evaluated, and treated.

Risk-taking is an important business strategy. A new design and technology often come with a certain level of risk. Most times, taking technical risks to advance business is inevitable, and requires multifaceted risk analysis. Here are two examples of risk inherent to technical challenges and to market uncertainty. One might consider how these given examples could be accounted for proactively, i.e. before the consequences came about.

- Foldable screen cell phones were introduced in the late 2000s as a "coming soon" technology, yet they did not enter the market until 2020, because of unforeseen technical issues beyond original concept development.
- A minivan vehicle model featured second row seats that swiveled to face the third row. This new seating design was introduced in 2008 but then dropped three years later because of lack of customer interest (AN 2010).

In both cases, the technical risks taken by these businesses were dynamic, requiring a broader scope of risk analysis than simply focusing on just product development, or just market uncertainty.

**Risk Treatment**
There are several risk treatment strategies. ISO 31000 has similar categories of dealing with risks (ISO 2018). Adopting a strategy or mixed strategies for a risk is an art form and a deliberate management decision.

1. <u>Avoidance:</u> To choose, if possible, not to take a particular risk, thus avoiding its costly and disruptive consequences. For example, if a new technology is not proven fully reliable, it may be a good idea to avoid the risk by not using

the technology. Deferring a risk or risk-associated work to a later time is a common practice.

2. <u>Reduction:</u> To mitigate the magnitude of a consequence by adjusting certain components of a business plan or process, or by reducing its overall scope. There are different ways to reduce risks, e.g. having a backup plan when trying a new technology. This way, the risk is still there, but its impacts can be reduced to acceptable levels. Another way to mitigate a risk is to reduce its likelihood of happening.

3. <u>Transference:</u> To transfer risky work (entirely or by portion) to a third party, for example, outsourcing a project, part, or business function to a contractor. This does not eliminate a risk, but it transfers responsibility at a cost. One example is the purchase of a health insurance policy in the healthcare sector.

4. <u>Sharing:</u> To distribute a risk to multiple organizations, departments, or individuals. Some risks may be too costly to absorb by one organization. For instance, to establish a partnership between two companies for the development of an autonomous vehicle.

5. <u>Retaining:</u> To make an informed decision to accept a risk and deal with its potential fallout. This is a common practice because it is unlikely to be successful without taking at least some risk.

Some predictable risks may be avoidable. However, risk avoidance also implies losing opportunities and jeopardizes future competitiveness. In addition, risk reduction may not always be feasible. "Prioritizing the risk management processes too highly could keep an organization from ever completing a project or even getting started" (Stamatis 2019). A risk-taking strategy built with components from the other four types of risk treatment speaks to the competence of senior management in taking risks at an acceptable level, and the progressive vision of those senior managers.

### 2.2.3 Risk Evaluation

**Concept of Risk Evaluation**

A foundation of risk management is a process called risk evaluation. Composed of various methods and tasks to assess risks that may jeopardize achieving a goal, risk evaluation can be conducted for different aspects of a given business. For example, a risk evaluation may be conducted to figure out the risks in supplier selection and management specifically.

The quality risk of a product or service may be defined in three or five levels. The following list is an example of product risk in three levels. If risks are categorized into five levels, they may be labeled: critical, high, moderate, low, and none. The actual definitions of the levels depend on applications, e.g. the healthcare service:

1. High risk if a product:
   - Has a possible product liability issues (e.g. crutches, insulin)
   - Has a critical feature for customers (e.g. pacemakers, oxygen systems)
   - Is to be built in a critical manufacturing process (e.g. surgical instruments)
   - Adopts a new, unproven technology
2. Moderate risk if a product:
   - Has a complex design (e.g. surgical robotics, prosthetics)
   - Impacts the quality of a final product as a main part (e.g. bandages in a trauma kit)
   - Has a high material, rework, or warranty cost
   - Is to affect customer perspectives (e.g. waiting times)
3. Low risk if a product:
   - Has no issue mentioned for high and moderate risks
   - Has a proven history of excellent quality
   - Uses standard or catalog parts

The high-risk issues should be addressed with the countermeasures based on the five treatment strategies already mentioned. The moderate (or medium) risk issues should be assessed to decide the further actions. Most times, the product/part should be closely monitored in operations if the issues cannot be resolved economically.

The severity levels of a risk impact are discipline specific. For a healthcare service, the severity of risks can be measured by the potential harm to patients in five levels:

1. <u>Negligible</u>: Only inconvenience or temporary discomfort to patients.
2. <u>Minor</u>: Likely causes patient temporary injury or impairment not requiring a professional medical intervention.
3. <u>Serious</u>: Likely causes patient injury or impairment requiring a professional medical intervention.
4. <u>Critical</u>: Likely causes patient permanent impairment injury.
5. <u>Catastrophic</u>: Likely causes patient death.

Another important factor is the likelihood or probability of risk occurring. For example, one may measure the likelihood in five levels: 1) rare, 2) unlikely, 3) possible, 4) likely, and 5) almost certain. The exact definitions of the five levels are case dependent. With the severity and likelihood of risks, a risk priority can be quantified for further considerations as:

$$\text{Risk Priority Index (RPI)} = \text{Severity} \times \text{Likelihood}$$

If both severity and likelihood are on a scale of five, the RPI is in a range of 1–25. A risk with RPI = 25 means it has both highest severity and likelihood. The risks with RPI > 15 should be treated seriously, with few exceptions.

**Risk Analysis**

Risk evaluation and analysis can be conducted quantitatively, qualitatively, or a mix of the two. A quantitative analysis is preferred if quantitative data are available. The likelihood and severity of a risk may be directly calculated based on relevant theories, such as statistics or the Monte Carlo simulation, as the situation warrants. Because of technical contents, a quantitative analysis in risk management may involve the use of specialized software by trained professionals. Each quantitative analysis method has its restrictions. For example, the Monte Carlo simulation only serves to estimate the overall probability of risk in a project, rather than for individual risks of that project.

A qualitative analysis may serve as a basis and lead into a quantitative analysis for an in-depth study. The common qualitative methods include:

- Team brainstorming
- Questionnaire surveys
- Structured interviews
- Focus group studies

In a risk management process (Figure 2.9), a review tracking form (Table 2.2) may be used as a quick reference for addressing the first three steps (identification, assessment, and planning) of risk management.

The output of a qualitative risk evaluation can be presented in a few two-dimensional matrices. For example:

1. The likelihood of a risk and the severity of its impact (or the RPI discussed earlier).
2. The expected benefits of a proposed change and the severity levels of risk impact.
3. The difficulty levels of risk treatment and the severity levels of risk impact.

The conceptual examples of the second and third matrices are shown in Figure 2.11 (a) and (b), respectively. The scales for the levels of benefit, severity

**Table 2.2** Risk review form sample of risk management

|  | 1. Identification | 2. Assessment | | | 3. Planning | | |
|---|---|---|---|---|---|---|---|
| No. | Risk | Impact | Likeli-hood | RPI | Treatment | Time | Respon-sibility |
| 7 | ...<br>A new feature<br>... | 3 | 3 | 9 | Mitigation (adjust design) | 5/18 | HW design |

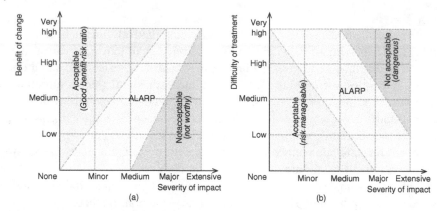

**Figure 2.11** Examples of risk assessment matrices.

of impact, and difficulty of risk treatment can be a number between zero and one, in five levels like in the figure, or a percentage range, etc. In the figure, there are three zones. The middle zone is sometimes called As Low As Reasonably Practicable (ALARP). Any risk in the ALARP zone requires analysis and reviews to find a practical way to reduce the risk probability, impact, etc., and decide its acceptability. These types of two-dimensional diagrams can be useful to compare alternatives visually.

In addition, a high risk may be evaluated twice: first as a preliminary study when a risk is identified; then again for the improvement when treatment measures are applied for the risk. Without good teamwork and brainstorming, avoidable redundancies produce unnecessary investigations, while other important risks are left uninvestigated entirely.

### 2.2.4   Event Tree, Fault Tree, and Bowtie Analysis

#### Event Tree Analysis
An event tree analysis (ETA) is a logical modeling technique that can be used for risk analysis. The concept of ETA is that an initiating (root) event is broken down into multilevel, smaller events in a complex system (refer to Figure 2.12). Quantitative methods, e.g. probability and statistical analysis, can be used in ETA.

For a risk analysis, the root event in an ETA is a specific risk or potential problem. Each path from the root event to the smaller events on a tree has a known probability. The likelihood relationship between the root and any small event risk can be calculated based on the probabilities of the events along the path. In the illustrated example, the calculated probability of a particular risk ($R_{32}$) is 7%

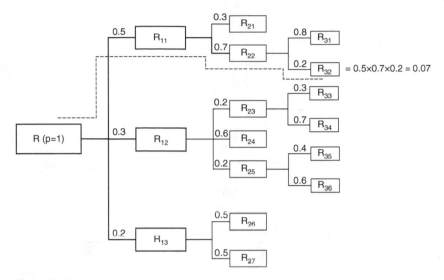

**Figure 2.12** Illustrative example of ETA.

of the root risk. Therefore, using ETA can help quantitatively identify a risk with its probability of occurrence.

The tree structure development can be achieved by team discussion and brainstorming, given probability data. An ETA can be conducted in four steps:

1. Define the root event
2. Develop the overall tree structure
3. Determine the probability of each event
4. Deduce the quantitative relationships

The challenging part here is usually Step 3, as quantitative data is needed for each event. This must be derived from measurements or observations. Quantitative data may come from interviews (Özfirat et al. 2017), theoretical analyses (Mineo et al. 2017), history data, simulations, and/or tests. The typical examples cited in ETA applications are for nuclear power plants and chemical plants. A general standard for an ETA is IEC 62502 (IEC 2010).

**Fault Tree Analysis**

Fault tree analysis (FTA) is a similar technique and can be used for risk management. In an FTA, each element is often called a fault. The principal purpose of an FTA is to identify potential causes of system failures and understand the conditions of a failure. An illustrative example of FTA is shown in Figure 2.13. In addition to quantitative failure probabilities, an FTA has a logic relationship.

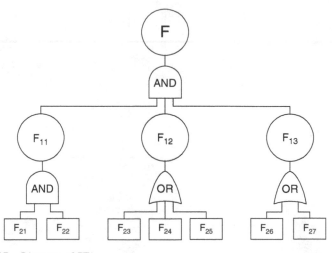

**Figure 2.13** Diagram of FTA.

One difference from an ETA is that an FTA uses the Boolean logic symbols (called gates) to describe the relationship between the elements at different levels. A Boolean logic symbol combines a series of lower-level elements into a higher-level element. Each gate has inputs and outputs and defines the relationship between them. Boolean logic gates include AND, OR, Priority AND, Exclusive AND, and Exclusive OR. For example, using an AND logic gate, its output is true only if all inputs are true.

The FTA approach has been standardized. For example, NASA uses NUREG–0492 (NASA 2002), the civil aerospace industry uses SAE ARP4761 (SAE 1996), and the military uses MIL–HDBK–338B (DOD 1998). For general industry use, it is recommended to use IEC standard IEC 61025 (IEC 2006).

FTA may be used for a causal relationship between multiple causes (inputs) and a failure or risk, while ETA is often used to investigate the consequences of a risk. These two approaches are complementary and can be used jointly in a single analysis. Examples of FTA applications include for the design reliability of an electromechanical system (Jiang et al. 2019) and the risk of fire in a gas processing plant (Hosseini et al. 2020).

**Bowtie Method**
Another network analysis method is called the bowtie (or butterfly diagram) method, named for its unique shape. A generic bowtie diagram (Figure 2.14) shows the relationships among a potential hazard, event, root causes, consequences, and measures (or controls).

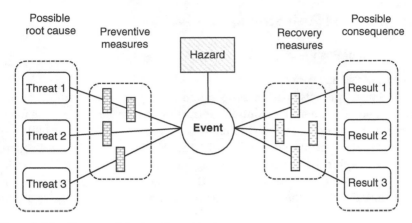

**Figure 2.14** Diagram of the bowtie method.

In a bowtie diagram, a "hazard" can be a danger or risky situation and an "event" is something that happens in the hazard situation. A "threat" on the left side of a bowtie diagram is a root cause for the event, while a "result" on the right side is a consequence of the event. The "measures" are the factors and constraints: the preventive measures and recovery measures are the controlling and influencing factors to the threats and results, respectively. In addition, there are often multiple measures displayed in the figure. During analysis, defining and quantifying measures can be challenging.

The left side of a quantitative bowtie diagram is like an FTA, while its right side functions like an ETA. A qualitative bowtie diagram closely resembles a cause–effect diagram (discussed in Chapter 8), with additional measures. A qualitative bowtie may be a good starting point for any new analysis. From there, a quantitative analysis can be introduced for each individual element of the resultant bowtie, using what methods are deemed appropriate for the given scenario.

The bowtie method has garnered popularity among modern risk management professionals, and its modular design makes it applicable to many situations. For example, applying a new technology to a product, one may start defining potential issues and eventually complete a bowtie diagram for the risk analysis of adopting that new technology. The bowtie method has been used in various industries, such as healthcare (McLeod and Bowie 2018) and the process industry (Aneziris et al. 2017). One issue related to implementing the bowtie method is based in the linear model of causality (Ruijter and Guldenmund 2016).

## 2.3 Pull and Push Strategies

### 2.3.1 Pull or Push

#### Concepts of Pull and Push

Pull and push concepts in business operations are a part of lean manufacturing planning and execution. The control information in a system is a driving force to operation execution. One way to understand the mode of push or pull is to compare the control information flow with the physical process flow in an operation. A simplified illustration of push and pull operations is shown in Figure 2.15 (a) and (b). In a push mode, the operation planning and execution are mainly driven by the market forecast and resources planning. In a pull mode, the execution of individual operations is authorized according to the system's status, the requests from downstream processes, and ultimately the customers. Therefore, pull operations are primarily customer driven, while push operations are essentially schedule driven.

In the push style of operations, a product producer or service provider pushes their products or services through their systems to consumers. Push operations are effective, but they are not driven by the demand of customers. Although push practitioners think they know what their customers want, they are less likely to exactly meet customer expectations. Thus, a typical issue of push

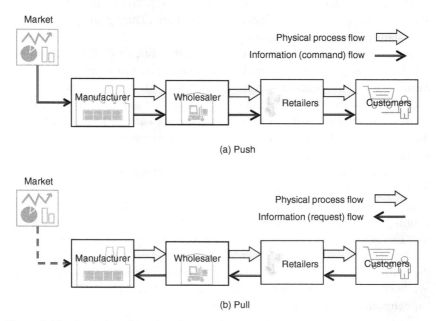

**Figure 2.15** Depictivepull and push scenarios in business operations.

operations is a high inventory (a type of waste). As a result, they often have to offer price discounts when demand wanes. The overproduction from push operations can also cause quality issues and slow responses to the market.

A pull mode of operations is about the control flow in the opposite direction shown in Figure 2.15. In a pull system, a downstream operation provides requests, such as what part and quantity, to its upstream operation. The upstream operation works accordingly. In other words, a pull operation is based on demands rather than forecasts. Many pull operations run like a supermarket. In such a pull system, each department and process has a predefined storage space that holds an amount of goods. The storage can be minimized with accurate and up to date demand information, allowing each department to simply replenish what is withdrawn from the supermarket. Whenever there is overstock, the operation has encountered a push mode.

These principles apply to services as well. As an example, reference the differences between broadcast television (TV) and TV streaming service provision. Traditional business operations, including cable TV and satellite TV, are mainly a push operation. Customers are forced to accept whatever programs are broadcast to them. They have limited choices in exactly what they receive, and usually the service comes with unwanted commercials. The providers feel that they understand what their customers want, but the system is built around efficiency of provision (cost effectiveness) of the service, for the benefit of the providers. In contrast, the relatively new TV streaming services represent a pull mode of operation. Here, the customers choose what and when to watch, and usually without unwanted additional features (commercials). Providers react to what the customers "pull" them to provide, and so they replenish a supermarket of that kind of TV content as necessary.

### Discussion of Pull and Push

If one has a monopoly business environment for a product or service, one can ignore the customers and listen only to the voice of a producer, forecasting "push" operations. Such a push mode and its habits may be a legacy from a past when a monopoly was enjoyed. That is, "push" is in place because it is easier to listen to the producer than to the customers. Therefore, a push strategy works well in a stable/consistent market. In addition, some products must be produced ahead of time based on predictions and resources planning, necessitating a push condition.

When market and customer demand constantly change, a pull organization can pull products or services to best suit the customer's wants. A pull scenario can be more cost effective with a minor inventory than a push scenario. However, it is challenging to manage every stage of a pull system throughout the operation, given the complexity and variety of requirements. A pull operation requires accurate and real-time information on system status, the requests from the downstream processes and customers, which can be challenging. Therefore, pull and push strategies are often jointly used for complex systems and operations.

Table 2.3 lists a general comparison between pull and push for business operations for discussion.

In principle, an effective business system should aim for a "pull" framework as a business operation is designed to add value, as defined by customers who can get what they want and when they want it. On the quality side, many approaches, such as quality function deployment (QFD) and FMEA, can support a pull strategy. These approaches are discussed in the following chapters.

### 2.3.2 Innovation-push

#### Arguments on Push and Pull

Many lean principle practitioners believe the pull scenario of operations is better than push in most cases because the pull focuses on customers, driving a business to be cost effective. A pull style or customer-driven business is likely to be more successful in market competition. Many organizations are well known for their good quality products and services, having already implemented the lean manufacturing and six-sigma principles.

However, applying lean manufacturing and six-sigma principles alone does not necessarily make a business grow. Another strategy, innovation-driven or technology-push growth, is playing an increasingly important role in today's market competition as many customers choose to try the latest interesting technologies. Product producers or service providers can gradually lose these customers, even with excellent products or services, if they

**Table 2.3** Characteristics of pull and push business operations

| Strategy | Push | Pull |
| --- | --- | --- |
| Environment | Monopoly or oligopoly | Global competition |
| Operation control | Produce in quantities linked to capacity | Produce what and when customers want |
| Focus | Based on the voice of producers | Based on the voice of customers |
| Driving force | Predict, design, and make to customer | Design and build on customer needs |
| Pricing | Set prices based on costs | Charge what the customer will pay |

fail to innovate. A *Harvard Business Review* survey study showed that both continuous improvement and innovation are equally important as business success factors (see Figure 2.3) (HBR 2013). One recent study identified innovation as the key to simultaneously increasing service quality and productivity (Rew et al. 2020).

The discussion on push and pull strategies may be extrapolated back to business strategy: in terms of the three key factors for business success (Figure 2.16). The first one is customer satisfaction, and the second is operational effectiveness. Both can be continuously improved to maintain a market share by implementing many approaches, including quality management systems (QMS), lean manufacturing, and six-sigma. The third key factor is the technical innovation effort, undertaken to draw customers and grow the market share.

For example, as a newcomer to the automotive market, Tesla has been facing tough competition from the existing automakers, including BMW, GM, and Toyota. Mainly due to its technical innovations, Tesla grew its electric car market share to 60% in the fourth quarter of 2019 from about 25% in the first quarter of 2013 (Lambert 2020). Tesla's success was significant: its US market volume of electric vehicles was 97,102 and 326,644 units (a 236.4% increase) for 2013 and 2019, respectively (DOE 2020).

**Innovation Driven**

Technical innovation is one of three cornerstones for a business success, as discussed. For long-term success, quality is a necessary condition, while innovation is a sufficient condition. In recent decades, innovation is more and more imperative for new products and services for existing companies, and in general for new startup companies.

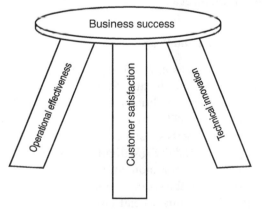

**Figure 2.16** Three cornerstones for business success.

To be innovative, a product producer or service provider should study future demands, do intensive research and development (R&D), and plan their novel products or services. Such an innovation process is in a push mode in nature, as a novel product or service is not expected by customers. Once a new technology or new application of an existing technology is developed, the producer or provider hopes to surprise and delight customers, reaping massive rewards in the marketplace, in exchange for the front-end resource investment. As new technologies develop, old ones pass off in obsolescence or obscurity. Two examples are the 35mm film cameras that lost market share to digital cameras, and physical keyboards in cell phones that lost to touch screens.

A foundation for an innovation-push strategy is R&D. Here are a few contemporary examples of "revolutionary" technologies, currently in R&D, that may go beyond the expectations of most customers:

- 3D-printing for residential house building
- Smart headsets to view manuals or error proofing in operations
- Autonomously driven vehicles
- Delivery drones

Clearly, such an innovation-push has inherent risks of technology, regulation, customer acceptance, etc. For example, many people are excited about autonomous vehicles, while others are concerned about driver and pedestrian safety and the impact on traditional personal transportation. Thus, risk assessment and management, discussed in the previous section, can be more important to innovations than they are to existing products.

The innovation-push concept has been practiced in various fields, e.g. policy study (Kim and Brown 2019), manufacturing application (Singla et al. 2018), and technology management (Sarja et al. 2017). In addition, it may be a good idea to integrate the customer-pull and innovation-push approaches for new service development (Geum et al. 2015).

### 2.3.3 Challenges to Pull and Push

Both pull and push strategies have their advantages and practical issues. One point from the discussion in this chapter is that the advantages and issues occur under certain conditions and environments. For example, Toyota Production System (TPS) "historically, has evolved under Toyota's singular conditions, and that its substance can be transferred to other structural contexts only with difficulty" (Williams et al. 1992). Hyundai revised the TPS in the development of a more suitable one for its own production systems. Hyundai rethought the standard approach, and came up with a system that was better fitted to its special culture and production standards (Lee and Jo 2007).

Here are several additional points for reflection and discussion. Challenges and/or drawbacks to a customer-pull strategy may include:

- The strategy may be impractical or uneconomic for many complex or long-lead products and services, as their operations are necessarily in pull mode. For example, a customer's special-order vehicle cannot be available in a few days (yet). It is proactive calculation, not a reactive response, that resolves demand for strains of seasonal vaccines, knowing the quantity needed in the marketplace.
- System readiness, e.g. the potential logistic issues associated with a long supply chain, is a concern for receiving materials on time for the final assembly of a product. Strong and specific contractual language addressing on-time delivery is important. Internal examples would relate to this delay between supporting departments and frontline representatives.
- Culture change inside an organization can cause a shift, from an engineering mindset to value creation mindset, even to customer-oriented mindset, and as such the values defined in an organization should live consistently across the operating systems (Anonymous 2011).
- Diversity and government regulations. For example, the Canadian government redefined cannabis from a drug to a food additive, while the European market still considers cannabis to be a drug.

Pushing a new product, technology, or service via innovation can be risky. Special attention and more research are needed. For example:

- R&D's alignment with what the customer wants. It is possible that the customer's interest differs from or even contradicts the company's direction.
- Additional administrative efforts to keep up with the customer's changing wants, needs, and desires. Effective communication with the customer demands that the departments inside an organization ensure the right things are always ready at the right time.
- Analyzing the everyday experiences of customers to gain that contextual value of those customers (Geum et al. 2015). One should integrate a technological push into the overall strategy.
- Short lifespan (or quick update) for new technologies. When a technology is new, a producer may push it to the market. However, the peak popularity and sales can be short. Innovation needs to be continuous.

## Summary

### Strategic Management

1 Strategic development includes benchmark analysis, customer, internal structure, assessment, goal setting, policies, and guidelines.

2 Abalanced scorecard includes four perspectives: customer, finance, internal process, and innovation and learning, which are all related to quality vision and strategy.

3 The SWOT analysis is an approach for strategy development.

4 Hoshin planning is an approach in strategic development with seven steps, emphasizing communication and employee engagement.

5 An X-matrix, resulting from Hoshin planning, comprises four quadrants: medium-term objectives, annual objectives, deployment, and improvement.

6 Catchball is a communication process used across different management levels.

### Risk Management and Analysis

7 Risk management is a process to approach identifying, evaluating, and developing a plan to handle potential problems and achieve goals.

8 A primary principle of risk management is to address uncertainties, along with 10 other principles.

9 The process steps of risk management are identification, assessment, planning, execution, monitoring, and review.

10 Typical risk treatment strategies include avoidance, reduction, transference, sharing, and retaining.

11 Risks are often categorized into three or five levels based on their severity of impact, difficulty of treatment, benefits of change, etc.

12 Various methods, e.g. ETA, FTA, and bowtie analysis, can be used for risk analysis. Quantitative analysis is preferred.

### Pull and Push Strategies

13 Pull and push operations can be viewed by the relationship between information flow and process flow.

14 Pull and push in business strategies have their own advantages and disadvantages.

15 Customer-pull is a way to ensure quality and effective operations, most times.

16 Innovation-push, or innovation-driven growth, plays an important role in market competition, with considerable risks and challenges.

## Exercises

### Review Questions

1 Discuss one or two quality aspects in a BEA with an example.

2 Explain a SWOT analysis with an example.

3 Review the results of the survey study (Figure 2.3), and discuss its findings.
4 Discuss a special characteristic of Hoshin planning.
5 There are seven steps in Hoshin planning. Select a step and discuss its application.
6 Find an application example for a catchball process.
7 Talk about a common issue in strategic planning, based on your observation or experience.
8 Identify a challenge in strategic planning, based on your observation or experience.
9 Discuss an element of risk management with an example.
10 Comment on incorporating PDCA in risk management.
11 Review a method of risk identification with an example.
12 Discuss a method of the risk treatment, e.g. reduction, share, or transfer, with an example.
13 Think of examples in work or life with high, moderate, and low risks.
14 Explain the probability application in ETA.
15 Explain the Boolean logic gates in FTA.
16 How do we recognize the operational styles of a business operation in push and pull?
17 Can both styles of push and pull be jointly used in an operation? Why or why not?
18 Discuss the significance of the three key factors for a business success.
19 Discuss a challenge and/or drawback to a customer-pull strategy.
20 Review the advantages and disadvantages of innovation-push strategy with an example.

## Mini-project Topics

1 Find an example of a BEA in a field and discuss the internal and external factors in the BEA.
2 Locate a paper on Hoshin application in your discipline, from a library database, or https://scholar.google.com, and summarize the major points.
3 Review the importance, and a roadblock to employee engagement in business operations, based on published literature.
4 Develop a simple Hoshin X-matrix for a new product or service.
5 Search a paper on risk management in your discipline from a library database, or https://scholar.google.com, to review the application.
6 The likelihood, severity, and benefits of a risk are considered in risk management. Develop a matrix to address these factors to evaluate the risk.
7 Review and provide your viewpoint on the statement: "Prioritizing a risk management process too high could keep an organization from ever completing a project or even getting started."

8 Search a paper on ETA, FTA, or bowtie analysis in your discipline from a library database, or https://scholar.google.com, to review the application.
9 Evaluate the statement: "Risk avoidance also implies losing opportunities and jeopardizes future competitiveness."
10 A survey showed continuous improvement and innovation are equally important. Do you agree or disagree with the conclusion?

## References

Akarsu, Z., Metin, O., Kuru, Y.Y. et al. (2018). Using adapted version of Hoshin matrix for selection of agile software development processes, *Proceedings of the 12th Turkish National Software Engineering Symposium*, Istanbul, Turkey, September 10–12, 2018.

AN. (2010). Swivel 'n. gone, *Automotive News*, November 29, 2010, https://www.autonews.com/article/20101129/OEM01/311299953/swivel-n-gone, accessed in August 2020.

Aneziris, O.N., Nivolianitou, Z., Konstandinidou, M. et al. (2017). A total safety management framework in case of a major hazards plant producing pesticides, *Safety Science*, Vol. 100, Part B, pp. 183–194, 10.1016/j.ssci.2017.03.021.

Anonymous. (2011). Nerves of steel in "glocal" challenge: ArcelorMittal's push and pull strategy, global focus, *Suppl. Excellence in Practice, Brussels*, Vol. 5, Iss. 3, pp. 13–16.

Arab, M.A., Stille, H., Johansson, F. et al. (2019). Developing a hospital disaster risk management evaluation model, *Risk Management and Healthcare Policy*, Vol. 12, pp. 287–296. 10.2147/RMHP.S215444.

Barnabè, F. and Giorgino, M.C. (2017). Practicing lean strategy: Hoshin Kanri and X-matrix in a healthcare-centered simulation, *The TQM Journal*, Vol. 29, Iss. 4, pp. 590–609. 10.1108/TQM-07-2016-0057.

Dahlgaard, J.J. (2019). Evolution and future of total quality management: Management control and organisational learning, *Total Quality Management & Business Excellence*, Vol. 30, Sup. 1, pp. S1–S16. 10.1080/14783363.2019.1665776.

DeRuntz, B., Kowalchuk, R., and Nicklow, J. (2014). Hoshin Kanri X-matrix drives engineering leadership program success, paper ID #10897, *121st ASEE Annual Conference & Exposition*, Indianapolis, IN, 15–18 June 2014.

DOD. (1998). Military Handbook – Electronic reliability design handbook, MIL-HDBK-338B, Defense Quality and Standardization Office, United States Department of Defense, Fort Belvoir, Virginia. https://www.navsea.navy.mil/Portals/103/Documents/NSWC_Crane/SD-18/Test%20Methods/MILHDBK338B.pdf, accessed in July 2020.

DOE. (2020). Maps and data – U.S. plug-in electric vehicle sales by model, United States Department of Energy, https://afdc.energy.gov/data/10567, accessed in July 2020.

Fonseca, L. and Ferro, R. (2016). A management trinity: Employee satisfaction, customer satisfaction and economic performance, *International Journal of Industrial Engineering and Management*, Vol.7, Iss.1, pp.25–30.

Ford. (n.d.). OUR PURPOSE, https://corporate.ford.com/about/purpose.html, accessed in August 2020.

Geum, Y., Jeon, H., and Lee H. (2015). Developing new smart services using integrated morphological analysis: Integration of the market-pull and technology-push approach, *Service Business*, Vol. 10, pp. 531–555. 10.1007/s11628-015-0281-2.

Gonzalez, M. (2019). Improving customer satisfaction of a healthcare facility: Reading the customers' needs, *Benchmarking: An International Journal*, Vol. 26, Iss. 3, pp. 854–870. 10.1108/BIJ-01-2017-0007.

Hanson, S.G., Scharfstein, D.S., and Sunderam, A. (2019). Social risk, fiscal risk, and the Portfolio of government programs, *The Review of Financial Studies*, Vol. 32, Iss. 6, pp. 2341–2382. 10.1093/rfs/hhy086.

HBR. (2013). The impact of employee engagement on performance, a Report by Harvard Business Review Analytic Services, https://hbr.org/sponsored/2016/04/the-impact-of-employee-engagement-on-performance, accessed in November 2019.

Hosseini, N., Givehchi, S., and Maknoon, R. (2020). Cost-based fire risk assessment in natural gas industry by means of fuzzy FTA and ETA, *Journal of Loss Prevention in the Process Industries*, Vol. 63, 104025. 10.1016/j.jlp.2019.104025.

IEC. (2006). *IEC 61025 Fault Tree Analysis (FTA)*, Geneva, Switzerland: International Electrotechnical Commission.

IEC. (2010). *IEC 62502 Analysis Techniques for Dependability – Event Tree Analysis (ETA)*, Geneva, Switzerland: IEC – International Electrotechnical Commission.

ISO. (2007). Committee draft of ISO 31000 risk management – Guidelines on principles and implementation of risk management, ISO/TMB/WG Risk Management, https://web.archive.org/web/20090325160441/http://www.nsai.ie/uploads/file/N047_Committee_Draft_of_ISO_31000.pdf, accessed in April 2020.

ISO. (2018). *ISO 31000:2018 Risk Management – Guidelines*, Geneva, Switzerland: International Organization for Standardization.

Jiang, Y., Gao, J.M., Sun, G. et al. (2019). Fault correlation analysis-based framework for reliability deployment of electromechanical system, The 2nd International Workshop on Materials Science and Mechanical Engineering, October 26–28, 2018, Qingdao, China, IOP Conf, *Series: Materials Science and Engineering*, Vol. 504. 10.1088/1757-899X/504/1/012113.

Kaplan, R.S. and Norton, D.P. (1992). The balanced scorecard – Measures that drive performance, *Harvard Business Review*, Vol. 70, Iss. 1, pp. 71–79.

Kim, Y.J. and Brown, M. (2019). Impact of domestic energy-efficiency policies on foreign innovation: The case of lighting technologies, *Energy Policy*, Vol.128, pp.539–552. 10.1016/j.enpol.2019.01.032.

King, B. (1989). *Hoshin planning: The developmental approach*, Goal QPC Inc., pp. 1–2, July 6, 1989.

Lambert, F. (2020). Tesla owns more than half the US market, keeps electric car sales growing, https://electrek.co/2020/02/04/tesla-electric-car-sales-us-market-share, February 4, 2020, accessed in July 2020.

Lee, B.-H. and Jo, H.-J. (2007). The mutation of the Toyota production system: Adapting the TPS at Hyundai motor company, *International Journal of Production Research*, Vol. 45, Iss. 16, pp. 3665–3679. 10.1080/00207540701223493.

McLeod, R. and Bowie, P. (2018). Bowtie Analysis as a prospective risk assessment technique in primary healthcare, *Policy and Practice in Health and Safety*, Vol. 16, Iss. 2, pp. 177–193. 10.1080/14773996.2018.1466460.

Mineo, S., Pappalardo, G., D'Urso, A. et al. (2017). Event tree analysis for rockfall risk assessment along a strategic mountainous transportation route, *Environmental Earth Science*, Vol. 76, p. 620. 10.1007/s12665-017-6958-1.

NASA. (2002). Vesely, W. et al. (2002). *Fault Tree Handbook with Aerospace Applications*. Vision 1.1, Washington, DC 20546: National Aeronautics and Space Administration.

Naude, M.J. (2017). A proposed operational risk management framework for small and medium enterprises, *South African Journal of Economic and Management Sciences*, Vol. 20, Iss. 1, 10.4102/sajems.v20i1.1621.

Nicholas, J. (2016). Hoshin Kanri and critical success factors in quality management and lean production, *Total Quality Management & Business Excellence*, Vol. 27, Iss. 3–4, pp. 250–264. 10.1080/14783363.2014.976938.

NIST. (2015). The Baldrige criteria 101, Baldrige Performance Excellence Program, The National Institute of Standards and Technology, https://www.nist.gov/baldrige/publications, revised 7/8/15, accessed in August 2020.

Özfirat, M.K., Özkan, E., Kahraman, B. et al. (2017). Integration of risk matrix and event tree analysis: A natural stone plant case, *Sādhanā (Indian Academy of Sciences)*, Vol. 42, pp. 1741–1749. 10.1007/s12046-017-0725-6

Rew, D., Jung, J., and Lovett, S. (2020). Examining the relationships between innovation, quality, productivity, and customer satisfaction in pure service companies, *The TQM Journal*, Vol. 33, Iss. 1, pp. 57–70. 10.1108/TQM-10-2019-0235.

Ruijter, A. de and Guldenmund, F. (2016). The bowtie method: A review, *Safety Science*, Vol. 88, pp. 211–218. 10.1016/j.ssci.2016.03.001.

Russ, W. (2013). *The Certified Manager of Quality/organizational Excellence Handbook*, 4th edition. Milwaukee, Wisconsin: Quality Press.

SAE. (1996). *ARP4761 Guidelines and Methods for Conducting the Safety Assessment Process on Civil Airborne Systems and Equipment.*Warrendale, PA: SAE International.

Sarja, J., Saukkonen, S., Liukkunen, K. et al. (2017). Developing technology pushed breakthroughs: An empirical study, *Journal of Technology Management & Innovation*, Vol. 12, Iss. 4, pp. 42–54. 10.4067/S0718-27242017000400005.

Silveira, W.G., Lima, E.P., Costa, S.E. et al. (2017). Guidelines for Hoshin Kanri implementation: Development and discussion, *Production Planning & Control*, Vol. 28, Iss. 10, pp. 843–859. 10.1080/09537287.2017.1325020.

Singla, A., Ahuja, I.S., and Sethi, A. (2018). Technology push and demand pull practices for achieving sustainable development in manufacturing industries", *Journal of Manufacturing Technology Management*, Vol. 29, Iss. 2, pp. 240–272. 10.1108/JMTM-07-2017-0138

Spross, J., Stille, H., Johansson F. et al. (2020). *Principles of Risk-Based Rock Engineering Design, Rock Mechanics and Rock Engineering*, Vol. 53, pp. 1129–1143. 10.1007/s00603-019-01962-x.

Stamatis, D.H. (2019). *Advanced Product Quality Planning*. Boca Raton, FL: CRC Press.

Tennant, C. and Roberts, P. (2001). Hoshin Kanri: Implementing the catchball process, *Long Range Planning*, Vol. 34, Iss. 3, pp. 287–308. 10.1016/S0024-6301(01)00039-5.

Vries, H.D. and Van der Poll, H.M. (2018). Cellular and organisational team formations for effective Lean transformations, *Production & Manufacturing Research*, Vol. 6, Iss. 1, pp. 284–307. 10.1080/21693277.2018.1509742.

Williams, K., Haslam, C., Williams, J. et al. (1992). Against lean production, *Economy and Society*, Vol. 21, Iss. 3, pp. 321–354. 10.1080/03085149200000016.

Wolter, J.S., Bock, D., Mackey, J. et al. (2019). Employee satisfaction trajectories and their effect on customer satisfaction and repatronage intentions, *Journal of the Academy of Marketing Science*, Vol. 47, pp. 815–836. 10.1007/s11747-019-00655-9.

Zameer, H., Wang, Y., Yasmeen, H. et al. (2018). Corporate image and customer satisfaction by virtue of employee engagement, *Human Systems Management*, Vol. 37, Iss. 2, pp. 233–248. 10.3233/HSM-17174.

# 3

# Customer-centric Planning

## 3.1 Goal: Design for Customer

### 3.1.1 Customer-driven Development

**Customer Satisfaction**

Although there are different benchmarks for customer satisfaction, a typical one is meeting customer expectations. Thus, customer satisfaction can be viewed as a status, and measurement, of customer feelings about a product, services, or both. In addition to being an outcome measurement, customer satisfaction may be viewed as a perceptual, evaluative, and psychological process. This broad spectrum of customer satisfaction requires in-depth studies to fully understand.

Customer satisfaction is an overall consumer behavior, and thus affected by various influencing factors. A single measurement may or may not provide reliable and comprehensive information. In addition, the data of customer satisfaction often comes directly from a set of customers. Most times, these data are indirect, e.g. sales records, profits, market share, customer complaints, etc.

Customer satisfaction is about the customers themselves, who received a product or service, and their influence on other potential customers. Positive word-of-mouth can translate into new customers. However, dissatisfied customers engaged in greater word-of-mouth discourse about their experiences than satisfied customers (Anderson 1998). Many studies have focused on this subject recently.

Customer studies are often called customer-satisfaction studies. Their aim is to understand what customers think, need, and want, and why they like or dislike a particular product or service. Such an understanding provides a foundation to develop a new product or service. The focuses of a customer study are on unsatisfactory aspects, reasons for being unsatisfied, and suggestions for improvements of a new product or service. For example, a customer study can

*Quality Planning and Assurance: Principles, Approaches, and Methods for Product and Service Development,* First Edition. Herman Tang.
© 2022 John Wiley & Sons, Inc. Published 2022 by John Wiley & Sons, Inc.

investigate possible gaps in the planning and execution of a product or service (Grigoroudis and Siskos 2009):

- Promotional gap: A product or service differing from the actually delivered.
- Understanding gap: Inaccurate perceptions of a product producer or service provider.
- Procedure gap: Customers' expectations not being translated into a product or service.
- Performance gap: Delivery different from the specifications of a product or service.
- Perception gap: Perceived by customers differing from the product or service actually provided.

Such gap-focused studies may reveal reasons behind given levels of customer satisfaction, and provide insight into opportunities for improvement.

### Customer Study Methods

As discussed in Chapter 1, a business development and operation should be end-customer focused. As Mr. Seddon saw it, a company strives on the idea that "finding out what matters to customers is central to the work of people who serve them" (Seddon 2005). There are various ways to get the information, including:

- Surveys, such as on customers, dealers, suppliers, employees, etc.
- Operational data, e.g. sales, warranty, complaints, returns, field service reports, etc.
- Other types of investigation, e.g. benchmarking, interview, focus group, mystery shopping, etc.

A commonly used method is a survey. Regardless of study method and data, the overall process of a customer study follows four steps (Figure 3.1):

1. Study planning (on the objectives, preliminary market/customer analysis, and scope of study)
2. Survey or method development (targets, questionnaire, sampling, and conduct plan, etc.)
3. Data analysis (qualitative and/or quantitative analysis, and interpretation)
4. Conclusion (reviews and suggestions)

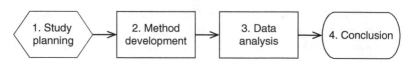

**Figure 3.1** Process flow of customer study.

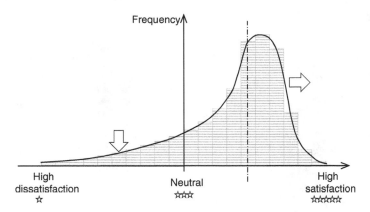

**Figure 3.2** Example of customer study result.

Depending on the study target, customer satisfaction can be based on overall or specific performance of a given product or service. Typical results of a customer study are shown in Figure 3.2.

For any business organization, its industry benchmark is an important reference. For an organization, product, or service, if it does not have the best level of customer satisfaction, a benchmarking study can show the gap between its performance and those of industry leaders.

There are several national or regional cross-industry measure indexes of customer satisfaction, such as the American Customer Satisfaction Index (ACSI), German Customer Satisfaction Barometer (GCSB), Swedish Customer Satisfaction Barometer (SCSB), and European Customer Satisfaction Index (ECSI). Figure 3.3 shows the 2020 ACSI scores of several industries (ASCI 2020). Note that the national ACSI score represents aggregate customer satisfaction across a broad swath of the US economy.

For a specific product, e.g. smartphones, or service, e.g. hotel chains, the ACSI provides customer satisfaction scores by brand. A business organization can compare their own customer satisfaction level to the ACSI or other index, find its standing in an industry, and develop new directions as necessary. Some researchers studied the relationship between the ACSI scores and financial performance of organizations (Schneider et al. 2009; Sun et al. 2013; Balvers et al. 2016).

**Overall Realization Process**
After finding customer needs and wants, the remaining question is how to address customers' demands to meet their expectations in product and service development. To convert customer demands to a new product or service, there

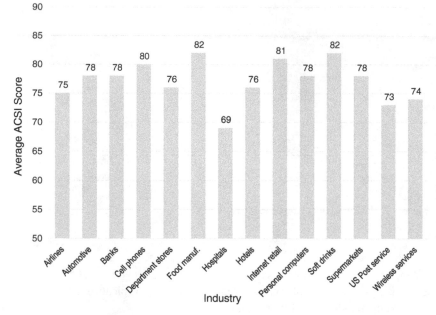

**Figure 3.3** 2020 ACSI scores of several industries.

are five generic steps, shown in Figure 3.4. Whether the process flow runs sequentially or partially simultaneously is an interesting question that will be discussed later.

0. An input from development work is <u>market analysis</u>, which is the understanding of customers' needs and market environment. The management team establishes a clear picture of who the customers are, what their interests are, and what the factors of their buying decisions are through market analysis. For example, online shopping has been extended to almost all types of goods, and shopping for cars online is a new customer demand.

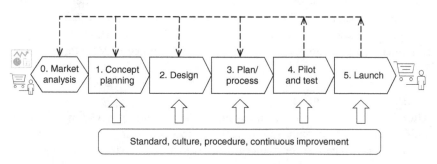

**Figure 3.4** Translation process flow of customers' demands.

1. The <u>concept planning</u> step addresses how to digest and convert ideas and information from market analysis into a workable plan. This planning process should be similar for all products or services. For a service, management sets the goals for key support metrics, like first response time and average resolution time. For a product, management set the goals of product functionality and performance. During the planning phase, management should consider the ever-changing expectations of customers, for new products or services to fit with these changes.

2. The <u>design</u> step is where the bulk of the work translates customer demands into a new product or service. The processes and requirements of design are discussed in depth in Chapter 4. In this step, technical challenges must be addressed and resolved. For example, new technology should be developed for online car shopping. Some £80 million was invested in the technology and startup for an online used cars shopping system called CAZOO (Sajid 2020).

3. This <u>plan/process</u> step is to realize the product or service designed, including all efforts to make the product or service ready and the coordination across departments. In this phase, the primary tasks are manufacturing process planning for a new product, or service development planning for a new service.

4. The <u>pilot and test</u> step builds a preliminary product model or service trial to test and validate its completion, functionality, and performance, before launching it to customers. This test step is vital to the success of a product or service. More detail is found in Chapter 4.

5. The last step is operational <u>launch</u>. For a service, this step may directly interface with end-customers. For a product, this step is to start a normal production, where the products are salable. Monitoring performance and continuous improvement is also important for this step.

During the development and realization process, the entire organization should share the same foundation on customer-centric culture, standards, procedures, innovation, and continuous improvement, which will be discussed further in the subsequent subsections and in Chapter 4.

### 3.1.2 Product/Process Characteristics

#### Types of Characteristics

A characteristic is a feature of a product or service, such as a size, shape, or color. For example, car acceleration is a performance feature for many car buyers, as a long-lasting battery is an important feature to smartphone buyers. As discussed in Chapter 1, in general, product quality often has eight dimensions, while service quality has five. All quality dimensions should be represented as product or service characteristics, addressed in development phases, and controlled in operational processes.

Normally, three levels are defined for the characteristics for a product or service, based on their significance to the product/service quality, and to customer satisfaction.

1. Critical: A critical product characteristic is related to consumer safety and/or in compliance with governmental regulation. This type of characteristic is sometimes called a critical-to-quality or critical-to-customer requirement. For example, the requirements on airbags and the performance of breaks of passenger vehicles are directly related to customer safety. In the pharmaceutical industry, the safety and compliance of a medicine must meet the Food and Drug Administration (FDA) requirements. In healthcare, a similar connotation is called a sentinel event, which is an unexpected patient safety event that results in death, serious physical or psychological injury, or the risk thereof.

2. Special, significant, or functional: A special characteristic of a product or service, directly related to the important functions, leads to customer satisfaction. For instance, the function of climate control in a car are a special characteristic. For a medicine, a special characteristic could be the controlled release of the drug substance. In healthcare, a near miss is as significant as an unplanned event that did not result in injury, illness, or damage, but had the potential to do so.

3. Standard or ordinary: Most characteristics of a product or service are called ordinary, as they are for the intended functions. To put it simply, standard characteristics equate to fitness for purpose. For instance, items such as a car radio, door locks, and cruise control could all be classified as standard characteristics. In a service, standard means regular work or basic functions.

### Discussion of Product Characteristics

Characteristic categories depend on a particular product or service. For example, cleanliness can be a standard or special characteristic for a service, e.g. hotel rooms, and is a critical characteristic for hospitals. Due to the complexity and variation of products and services, the interpretation of three levels of significance may also be different to the producers and providers within an industry.

To a specific product, the relationship between its characteristics and quality dimensions can be established in a matrix format, as exemplified in Table 3.1. Within a dimension, there may be several product features. Therefore, product features in one dimension can also belong to two categories of characteristics. In

**Table 3.1** Quality dimensions and product characteristics

| Characteristic / Dimension | Critical | Special | Standard |
|---|---|---|---|
| Performance | | √ | √ |
| Safety | √ | | |
| Features | | √ | √ |
| Reliability | √ | √ | |
| Durability | | √ | √ |
| Esthetics | | √ | |
| Conformance | | √ | |
| Serviceability | | | √ |

addition, the relation matrix should be built based on quantitative data analysis and team consensus as much as possible, to reduce the subjectivity in the relationship study.

Unique characteristics have different levels of impact on functionality. To address the differences, one may assign weight scores, such as 20, 10, 5, and 1, to individual characteristics (Table 3.2). By doing this, an overall quality demerit can be established for an entire product, e.g. a car. Alternatively, the weight scores can be based on how likely it is that customers will discover defects. For example, the weight score of a quality issue is assigned 50 as 50% of customers would notice it.

Because of the impacts of these characteristics on customer satisfaction, one should have different assurance plans in place during development and operations. All critical and most special characteristics of a product or service should have control and monitoring plans, and be studied by using failure modes and effects analysis (FMEA) and control plans (Lorenzi and Ferreira 2017).

**Table 3.2** Product characteristics and process quality assurance.

| Characteristic | Critical | Special | Standard |
|---|---|---|---|
| Weight score on demerit | 20 | 10 | 5 or 1 |
| In-process monitor | yes | yes | maybe |
| Final inspection/audit | yes | most of them | maybe |
| Acceptability | | | most of them |
| Action required | yes | most of them | |

**Discussion of Process Characteristics**

For a product, its manufacturing processes are used to realize the product and assure quality. In addition to product characteristics, some characteristics in operations can be identified as process characteristics. For example, the prevention and detection functions of quality issues in a manufacturing process are critical characteristics of that process. Quick repairability of defective products is another important capability of a manufacturing process.

For success in a business operation, product and process are about equally important. For example, the National Institute of Standards and Technology (NIST) has the Baldrige Performance Excellence Award (NIST n.d.). The award evaluation is 55% based on process and 45% on products/results. The Baldrige program assesses approaches, their deployment, learning and improvement, and integration between processes and operations. On products and results, it evaluates performance level, trend over time, competition comparison, and business integration among processes and results.

Process characteristics in manufacturing can be one of two types:

1. Directly affecting the quality or characteristics of the products to be built.
2. Indicating the manufacturing operations effectiveness, such as cycle time and uptime.

For the latter, indicators often combine a few key characteristics to represent the performance of an operation. For example, overall equipment effectiveness (OEE) combines three process characteristics: operational availability, process rate, and product quality (Tang 2019).

For product quality in a mass production environment, product design plays a critical role, determining most quality issues during a new product launch. Other realization efforts, such as inspections and continuous improvement, play critical roles.

For services, they themselves, e.g. healthcare, are processes in nature and highly involve customers. Therefore, one of the unique service characteristics is its direct relationship with customer feeling and interaction. Some types of services involve products, e.g. a banking service. In such cases, the service process characteristics and service product characteristics closely affect each other. One may ask, "Are we providing the customers with what they want?" and consider whether a service process or process feature can be changed. Service process characteristics can be more dynamic than manufacturing process characteristics; service process characteristics likely have immediate opportunities to improve.

## 3.2 Quality Category to Customer

Quality of a product or service can be viewed from different angles, and has the several dimensions discussed in Chapter 1. In the same way, customer satisfaction has multiple aspects, and too is affected by various factors. Many researchers have studied the differences between quality and customer satisfaction (Sherbini et al. 2017; Chen et al. 2020). Thus, it is important in practice to establish clear and accurate connections between quality and customer satisfaction.

### 3.2.1 Must-be Quality and Attractive Quality

**Must-be Quality**

Meeting customer's basic expectations is the fundamental requirement of a product or service. Such quality attributes of customer expectations are often called *must-be quality* (MBQ) or must-have quality, as customers simply assume that these quality attributes or features will be there, and take it for granted when they are fulfilled. For service quality, the cleanliness of hotel rooms is an example of MBQ.

Customers normally feel neutral for such type of quality attributes, as a good MBQ attribute is expected. When an MBQ attribute is not fulfilled, customers are dissatisfied with the product or service. As MBQ attributes are expected, their improvement may increase customer satisfaction, but not by much. However, if an MBQ attribute is lower than customer expectation, it can hurt customer satisfaction and sales. Therefore, one can consider MBQ as a foundation and necessary condition that a product or service can stay in a competitive market.

For example, an MBQ attribute for a new car is problem-free, or no major quality issues, for the first few years. Problem-free can be challenging for most products and services. For example, for passenger vehicles, J.D. Power does a survey every year for new vehicles, and measures their initial quality in terms of problems per 100 vehicles. J.D. Power's 2020 study based on 87,282 responses showed that the top five brands were Dodge, Kia, Chevrolet, Ram, and Genesis among 31 brands (J.D. Power 2020a). A major task for all automakers is continuously striving to reduce quality issues. So is the case for almost all product producers and service providers.

A recent research showed that the term MBQ might not be an accurate translation from the original meaning in Japanese (Horton and Goers 2019). A more accurate translation has been suggested, like "natural," "normal," or "taken for granted," to reflect the nature of MBQ. Alternatively, one may also call MBQ "expected quality."

**Attractive Quality**

While meeting customer expectations or conforming to specifications may be adequate to maintain a customer body, determining how to "delight" is an important pursuit and a sufficient condition for competitiveness. Some quality attributes of a product or service can pleasantly surprise customers – this is called *attractive quality* (AQ), "exciting quality," or a delighting feature. An AQ attribute may make customers amazingly satisfied, but does not cause dissatisfaction when not fulfilled, because this type of attribute is not expected. For example, an autonomous feature of a new car may make car buyers excited. On service AQ, an airline offering a free upgrade of seat class may be an example. A hotel may thrill their guests by providing a free special wine.

AQ can be a sales booster that provides a competitive market advantage. For example, conventional seating in passenger vehicles is fixed in place with only minor position and tilt adjustments. DaimlerChrysler created an innovative new seating system that can be folded completely into an under-floor compartment, which they introduced (US patent 6955386) in a minivan in 2005. This novel seating delighted vehicle buyers and boosted the minivan sales by 42% in 2005 compared with the previous year (Mayne 2007). Now, foldable seating has been adopted in many minivans, SUVs, and utility vehicles by most automakers.

AQ attributes normally come with new technologies and/or innovative designs, for example, newer models of cell phones and vehicles come with an increasing number of advanced features. J.D. Power conducts an annual study on the tech experience index (TXI) for vehicles. The 2020 TXI showed that the top five vehicle brands were Volvo, BMW, Cadillac, Mercedes-Benz, and Genesis among 31 brands, excluding Tesla, based on 82,527 responses (J.D. Power 2020 b).

It is interesting to note that an AQ attribute (e.g. feature, function, style, etc.) can become an MBQ attribute within a few years or even months from its introduction to the marketplace (Löfgren et al. 2011). Having started as an attractor, an AQ can soon become an MBQ since consumers are always demanding newer features and functions. There are opportunities to study AQ systematically, as AQ has not yet been fully studied in most areas of product and service development.

**Performance Quality**

Based on their impacts on customer satisfaction, quality attributes can be categorized, e.g. MBQ and AQ as just discussed. Some quality attributes of a product or service are proportionally associated with customer satisfaction, in the mode of "the more, the better" or called *performance quality* (PQ). The maximum driving range of an electric car and the fuel economy of a conventional car are the examples of a PQ attribute. Because of its approximately linear relationship to customer satisfaction, a PQ attribute is also called a one-dimensional quality. From the clear relationship of a PQ attribute to customer satisfaction, it would not be difficult to decide on and prioritize the investments on a PQ attribute in development. The payoff of investment and effort may be directly visible and measurable.

The sensitivity of PQ is an interesting topic. A PQ attribute to customer satisfaction can be modeled as a linear function:

$$y = ax + b$$

where $y$ is customer satisfaction, $x$ is the PQ attribute or feature of a product or service, $a$ is the sensitivity coefficient of the feature $x$ or the slope of the straight line, and $b$ is a constant.

In the relationship, a key is the sensitivity coefficient $a$ that determines the influence of the attribute of a product or service. When $a \approx 0$, the quality attribute has no influence on customer satisfaction with its performance. The value of the quality attribute can be challenged, and the attribute is called irrelevant quality. If $a < 0$, then the attribute of a product or service is perceived negatively by customers (Figure 3.5). The legroom of a car can be an example of a reverse PQ attribute, as the smaller the legroom is, the lower the satisfaction of car owners. Interestingly, the concept of PQ may be used to describe MBQ and AQ for

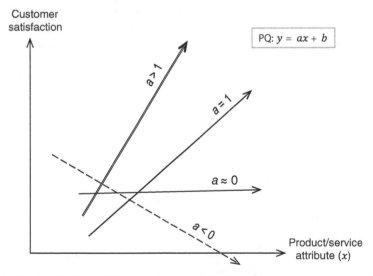

**Figure 3.5** Sensitivities of PQ to customer satisfaction.

their influences as well. If a quality attribute has no influence on customer satisfaction, its value can be challenged and called irrelevant quality.

Categorizing the quality attributes of a product or service into MBQ, AQ, and PQ establishes a meaningful perspective on their influences over customers, and helps the customer-centric development of a product or service.

### 3.2.2 Kano Model

#### Understanding the Kano Model

A product or service can have different levels of customer satisfaction, referring to Figure 3.6. The Kano model is an approach to link the two dimensions (product/service quality and customer satisfaction) conceptually and visually.

The Kano model was initially built by Noriaki Kano and his colleagues (Kano et al. 1984). Professor Kano promoted a two-dimensional quality model, integrating the quality of a product or service and customer satisfaction. There are several versions of the Kano model; a typical one is shown in Figure 3.7.

In a Kano model, there are two curves presenting MBQ and AQ. Shown in the figure, the improvements of MBQ and AQ affect customer satisfaction differently. In addition, a straight line presents PQ.

A quality attribute of a product or service is not necessarily represented by only one of the three quality types. For example, the wait time in a service may be an example of PQ – the shorter, the better. At the same time, a reasonable wait time, e.g. less than 10 minutes, may be also considered an MBQ attribute, as no satisfaction difference is noticed by customers in waiting five or seven minutes. Therefore, a quality attribute may be located somewhere between MBQ and AQ curves, rather than on a curve in a Kano model.

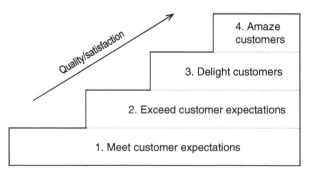

**Figure 3.6** Levels of quality performance and customer satisfaction.

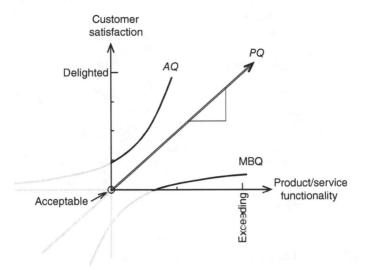

**Figure 3.7** Quality–customer Kano model.

In a Kano model, the horizontal axis is product or service functionality, which may be measured based on design specifications. Sometimes, the horizontal axis can be configured as investment. From the shapes of the three curves, innovative features to delight customers can be the most valuable to a product producer or service provider. If the three quality types (curves in a Kano model) can be quantified, the Kano model is more useful to prioritize development investments and efforts.

The vertical axis is customer satisfaction and excitement, which may be designed in five levels for a product or service. The five levels can be: 1) disliked, 2) acceptable, 3) expected, 4) liked, and 5) delighted. The scale should be used in line with a customer satisfaction survey.

### Applications of the Kano Model

As discussed, Kano modeling is to recognize the different influences of quality attributes of a product and service on customer satisfaction. An overall application process of Kano modeling comprises three steps, as shown in Figure 3.8.

1. To define the quality dimensions for a product and/or service, which can vary from the common ones discussed.
2. To identify the quality attributes with their respective quality dimensions.
3. To build a Kano model with the relations between the attributes and the three Kano quality categories.

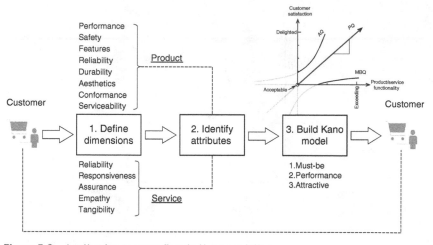

**Figure 3.8** Application process flow in Kano modeling.

In the modeling process, one may use a table to map the relationship between quality attributes and quality dimensions, referring to an example in Table 3.3.

To meet an essential requirement, the quality attributes of a product or service should be in the first quadrant of a Kano model. With a clear understanding of the three categories in Kano quality, one can determine the relative importance of dimensions, to better prioritize development efforts and investments on their respective attributes.

The Kano model has been used in product and service development and quality management, to support customer-centric development. The model is also an active research topic; Google Scholar delivered 13,300 results for the "Kano model" as a keyword search in March 2021. Application examples include vehicle profile (Yadav et al. 2017), product feature benchmarking (Gangurde and Pati 2018), project management (Lo et al. 2017), software design (Xu et al. 2019), pharmaceutical logistics (Chen et al. 2019), and online electronic services (Kurniawan 2017; Choi and Kim 2018).

Applying the Kano model remains challenging. One issue is that the definitions of a Kano scale are case dependent, which leads to inconsistent results (MacDonald et al. 2006; Shahin et al. 2013). For survey studies, the understanding of individual respondents on the descriptions of quality attributes may not be the same ones that affect evaluation results (Mikulic and Prebežac 2011). Therefore, the Kano model can serve as a guiding principle, but requires more studies. Grapentine (2015) suggested relying more on qualitative research to understand the nature of consumer decisions, and using quantitative analysis

**Table 3.3** Kano modeling example of hotel quality.

| 1. Dimension | 2. Attribute | 3. Quality category | | |
| --- | --- | --- | --- | --- |
| | | Must-be | Performance | Attractive |
| Reliability | • Accurate billing | √ | | |
| | • Wi-Fi/internet | √ | | |
| | • ... | | | |
| Responsiveness | • Check in/out | √ | √ | |
| | • Quick response | √ | | |
| | • ... | | | |
| Assurance | • Free breakfast | | √ | √ |
| | • Personalized notes | | | √ |
| | • ... | | | |
| Empathy | • Staff attitude | | √ | √ |
| | • Flexibility | | √ | √ |
| | • ... | | | |
| Tangibility | • Clean and comfort | √ | √ | |
| | • Parking security | √ | | |
| | • ... | | | |

for product evaluation processes. He also suggested conducting monadic tests of different product profiles to examine how changing one attribute affects consumer brand preferences and attitudes. There are yet more opportunities to study the Keno model and their applications.

# 3.3 Quality Function Deployment

## 3.3.1 Principle of QFD

### Development of QFD

Quality function deployment (QFD) is a proactive, customer-driven planning approach to building a connection between customer expectation, product design, and market competition for a product. The principle and process of QFD can apply to service development, too. QFD was initially developed in Japan in the late 1960s and is standardized in ISO 16355-1 (ISO 2021).

Applying QFD is a process – to build a three-factor relationship in a two-dimensional diagram, which is sometimes called the house of quality. There are a few different configurations of QFD, and a typical QFD structure shown in Figure 3.9. One may go through four segments to build a house of quality.

1. Identify and understand <u>what</u> customers expect (left side); the tasks include:
   - Collecting customer information
   - Recognizing customer expectations, which should be measurable
   - Assessing the relative importance of each expectation, and assigning weight scores
2. Evaluate product/service characteristics on <u>how</u> they contribute to customer expectations and correlate with each other (ceiling/top side):
   - For a product, evaluate its functions, features, and performance
   - For a service, evaluate its offerings and delivery
   - Analyze the correlation between the characteristics
   - Determine the improvement directions

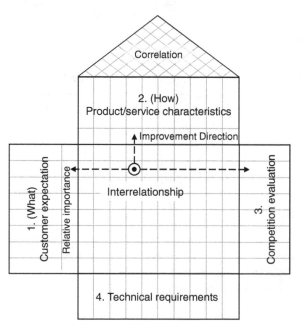

**Figure 3.9** Typical QFD structure with four segments.

3. Study market and rivals to know the <u>competition</u> (right side):
   - Identify the strengths of the product/service to specific customer expectations
   - Assess the competitors' positions, relative to the strengths of the product/service
   - Know how well the position of the product/service in the market is
4. Specify main <u>technical</u> specifications to be addressed in development (bottom) The items of this segment are directly in line with the items of the second segment:
   - Show technical data and/or targets
   - Determine the levels of difficulty and prioritization

With the four segments built, the next step is to establish the <u>interrelationships</u> among the first three segments by filling the centerpiece of the house. These interrelationships should show how strong they are, on a scale of 1–5 or 1–9. One should focus on customer expectations and the product/service characteristics by asking a few questions, for example:

1. Is there an interrelationship among the items in the three segments?
2. How strong is the interrelationship (in a 5- or 9-level scale)?
3. Which characteristics have a priority (based on the weighted customer data and interrelationship scales)?

### Considerations in QFD Development

After multiple iterations of cross-functional brainstorming and review, a cross-functional team can build the entire structure of the house of quality. Cross-functional teams taking part in the brainstorming process of QFD development is equally important as creating the QFD diagram. With intensive discussion, a team can reach consensus on the resultant elements.

From a QFD analysis, one learns a clear direction to take for a new product or service with justifications for customers. In addition, introducing and adopting a QFD process is not only a technical task but also leads to a mindset or culture change in a given business. A literature review showed that the benefits of applying QFD led to a deeper understanding of customers' needs and wants, better identification of opportunities for process improvement, more effective systemic thinking approaches, better communication, and a more transparent process in an organization (Gremyr and Raharjo 2013).

The resultant relationship in a QFD is qualitative, with a certain level of subjectivity. In a QFD development process, some other approaches can improve QFD validity, for example, using a focus group with a cross-department team and/or using a survey to help validate QFD. With other approaches and data, one can enhance input to QFD, improve its value, and promote consensus among diverse participants to a resultant QFD matrix.

Objectively, QFD itself may have debatable points. For example, can the inter-relationship between customer expectations and product/service characteristics be precisely presented with a symbol or a scale? The interrelationship can be nonlinear. The complexity of QFD interrelationships makes it difficult to provide completely objective results. Nowadays, customer demands are so diversified that it may or may not be appropriate to aggregate individual customer preferences into a collective indication or present them as an average number. Such aggregating and averaging can cause loss of useful information because of the over-simplified representations. Improving the effectiveness of QFD development and execution remains a concern of practitioners.

### 3.3.2 QFD Applications

**A QFD Example**

Figure 3.10 shows a QFD practice on a product called continuous tow carbon fiber (Jennings 2019). For context, continuous tow carbon fiber is a material used to weave carbon fiber-based products. In this example, the three key factors are: 1) customer expectations, 2) product characteristics, and 3) competition evaluation. With the QFD application, the interrelationship of the key factors are established on a scale of one to five. The details of the original competition evaluation are removed from the example.

In this example, the three-direction interrelationship is built in the center-piece of QFD. A weighted score is resolved for each attribute expectation of customers and each characteristic of the product. For example, "Affordable" attribute is $15 \times (1+1+5+5) = 180$ based on its importance weight score (15) and its interrelationship with the product characteristics of "Tensile strength" (1), "Modulus" (1), "Cost" (5), and "Spool weight" (5). The weighted scores for product characteristics are calculated in the same way. For example, "Tensile strength" is important to customers, as $15 \times 1 + 25 \times 5 = 140$ because of the "Affordability" (15) and "Within specification" (25).

The relative ranking of all product characteristics is available after comparing their weighted scores. For this example, the most important characteristic is the "Spool weight" as its weighted score is 250, which affects five out of the eight customer expectations. The QFD segment 4 shows that it is easy to do because of the difficulty level of "1." In contrast, the second most important product characteristic "Folds" (with a weighted score of 185) is technically difficult (level 8 in a scale of 1–10 in segment 4). This characteristic strongly (at the highest level 5 on interrelationship) affects two customer expectations that the company is not very strong (at level 3 out of 5 in segment 3) compared to its rivals. Therefore, this product characteristic should take priority for quality assurance.

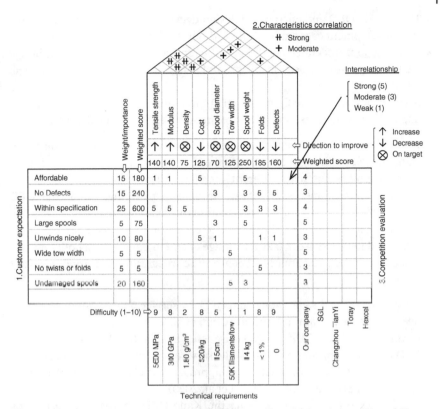

**Figure 3.10** Example of product QFD. *Source:* Jennings, J. (2019). Carbon Fiber QFD, QUAL 552 Quality Planning (CRN 13393), Eastern Michigan University, September 13, 2019.

Similar reviews can be conducted for other characteristics and eventually conclude with a to-do list with priorities for quality planning and execution.

### Additional Applications

The aforementioned example shows that a QFD can create an overall picture of quality status for a product with multiple facets. The QFD approach is suitable and beneficial for a new product or service and complex problem solving for an existing product or service. Here are some additional application examples across industries:

- Quality planning: Fostering alliances with customers for the sustainable product creation (Leber et al. 2018).
- Product design: Application of fuzzy QFD for sustainable design of consumer electronics products: an event study (Vinodh et al. 2017).

- Healthcare: QFD and operational design decisions – a healthcare infrastructure development case study (Dehe and Bamford 2017).
- Construction: Integrated condition rating and forecasting method for bridge decks using visual inspection and ground penetrating radar (Alsharqawi et al. 2018).
- Service: Developing the C-shaped QFD three-dimensional matrix for service applications with a case study in banking services(Shahin et al. 2018).

In addition, the QFD approach has become increasingly involved in decision-making processes (Serugga 2020). Both the Kano model and QFD address quality for customers but with different emphases. The Kano model addresses influences on customer satisfaction for quality planning, while the QFD approach identifies and prioritizes customers' expectations, while building interrelationships between those expectations and product characteristics. Therefore, it makes sense to take advantage of and integrate them both for quality planning. Much research has been on the integration and applications of both approaches:

- Integrating refined Kano model and QFD for service quality improvement in healthy fast-food chain restaurants (Chen et al. 2018).
- Integration of SERVQUAL model (discussed in Chapter 1), Kano model, and QFD to design improvement on public service system (Mansur et al. 2019).
- QFD and fuzzy Kano model-based approach for classification of esthetic attributes of SUV car profile (Avikal et al. 2020).
- Benchmark product features using the Kano-QFD approach: a case study (Gangurde and Patil 2018).

### 3.3.3 More Discussion of QFD

#### Multiple Levels of QFD

For a complex product, like a passenger vehicle, a single QFD may not be sufficient. QFDs may apply for main subassemblies (e.g. the infotainment system of a car) individually. A QFD study can be incorporated into the multiple phases of a product or service development, referring to Figure 3.11, to connect the customer expectations to the product and process characteristics. In such cases, the third factor, competition evaluation in QFD, may be optional.

After applying QFD into the multiple development phases, the main customer expectations are then translated into the execution of a service or the production of a product; customer expectations stay in the development process; and the link between customer expectations and product and process characteristics is established. Figure 3.12 shows an example of such interrelationships. This is also of interest to execution personnel, as they want to be fully aware how they work toward the expectations of their ultimate customers.

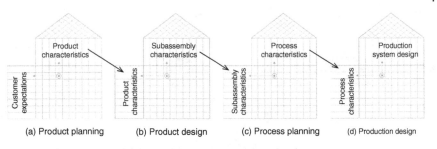

**Figure 3.11** Multi-level QFD application into product development.

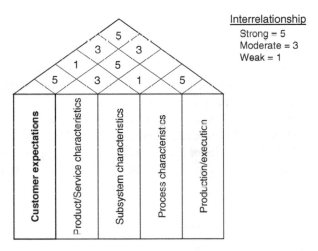

**Figure 3.12** Links among multilevel QFD applications.

Some research has been conducted on multilevel, or other complex forms of QFD. One study derived product characteristics into more detailed identities, in order to be easily recognized (Jia et al. 2016). Another study looked at the translation customer requirements into the design of product characteristics and manufacturing processes, to satisfy customers and minimize potential failure costs (Chen 2010). For a complex product, a QFD network was developed and used as a design framework during the early design stages (Li et al. 2015).

For a service, Koprivica et al. (2019) proposed a three-phase plan on the quality improvement of a hotel service. The three phases were: 1) service planning, 2) process control characteristics, and 3) action plans, as shown in Figure 3.13. This service example also showed the connection between the development phases. After the third QFD, the performance measurement of action plans could be conducted against the customer requirements in the first QFD.

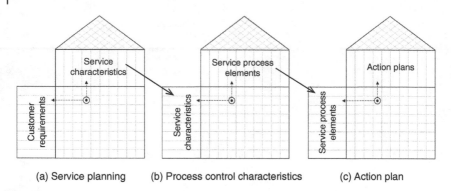

(a) Service planning    (b) Process control characteristics    (c) Action plan

**Figure 3.13** Multi-level QFD application into service development.

### QFD with Other Methods

In addition to the Kano model, QFD can be used in conjunction with other methods. For instance, QFD may be used in Hoshin strategic planning. As shown in Figure 3.14, QFD can apply across the multiple levels of Hoshin planning, to address different levels of customer needs and business activities. The application examples of joint QFD and Hoshin approaches include the customer satisfaction of a healthcare facility (Gonzalez 2019), managing

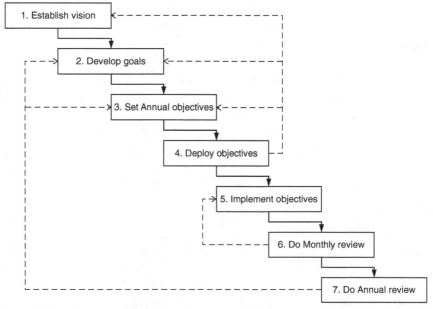

**Figure 3.14** Integration of Hoshin planning and QFD.

service quality (Pun et al. 2000), and student expectations in education (Gonzalez et al. 2019).

Another example of the integrative application is with FMEA, which is discussed further in Chapter 5. The outputs of QFD can be treated as an input to design-FMEAs and process-FMEAs. In translating customer expectations into quality characteristics using QFD, one study integrated the SERVQUAL model to set requisite success factors to improve quality in a private healthcare setting (Camgöz-Akdağ et al. 2013). Buttigieg et al. (2016) combined QFD with a logical framework approach in healthcare. Other integrative applications with QFD are in complex product redesign (Ma et al. 2019), quality service improvement (Altuntas and Kansu 2019), and failure analysis (Shaker et al. 2019), etc.

One may need to consider the characteristics and limitations of QFD (Poel 2007; Ginting et al. 2018). As a qualitative approach, QFD applications can be somewhat subjective. Introducing new technologies, and subsequent validation, may improve QFD reliability and validity. Many researchers have been incorporating fuzzy logic principles with QFD (Sivasamy et al. 2016). Others have integrated artificial intelligence techniques, such as ANP and AHP, with QFD and fuzzy logic (Zaim et al. 2014). To practitioners, these new technologies and different approaches should be embedded into QFD software for integrative applications.

## 3.4 Affective Engineering

### 3.4.1 Introduction to Affective Engineering

#### Concept of Affective Engineering

Affective engineering (AE) is a relatively new quality planning approach that guides professionals to transform customers' feelings and perceptions into executable design parameters and actions. AE was created by Dr. Mitsuo Nagamachi who called it Kansei engineering (Nagamachi 2001). *Kansei* is a Japanese word that means sensitive, emotional, or affective in English.

Dr. Nagamachi believes that sensation, perception, and cognition are separate processes. This is congruent with a psychological theory that suggests humans operate in three domains: physical, cognitive (rational thinking), and affective or attitudinal. Product producers tend to operate in the area that can be most readily measured: cognitive and sensory (what humans can perceive and measure in the physical world). Focal to AE is the notion that consumers also operate in the perceptual/attitudinal domain; one that is difficult to define and measure, and may overlap with the other two domains.

Cognition is a human sense, but not addressed much in most quality management systems. Perhaps this is because quality was historically defined by statisticians/engineers (such as Shewhart, Juran, Deming, and Taguchi) and many quality professionals worked in discrete part manufacturing, where measurement is relatively straightforward. Cognition measurement can be more challenging with customers' perceptions (or from psychology, an affective domain), which play a key role in how the delivery of a product or service is perceived.

A product producer or service provider considers the features of new goods or services from a customer's standpoint, and figures out how to specify, measure, and monitor the features during development. Traditional quality methods have much to do with using design specifications to meet customer physical requirements, while AE acknowledges the cognitive biases of a producer or provider and attempts to accommodate the affective/perceptual/attitudinal reactions of customers into product or service development.

To delight customers, one must first think about and be able to measure what they want, in addition to their needs, in both physical and perceptual terms. The perceptual inputs may include the items that customers can see, hear, smell, feel, taste, etc. From this point, AE is an improvement over conventional methods that focus solely on the physical requirements of customers (see Figure 3.15). In quality planning, quality professionals closely and collectively work with development management for both physical and perceptual aspects of quality for customers.

**Overall Process of AE**

Applying AE is a novel approach to product and service design. Dr. Nagamachi initially considered six types of AE methods, listed in Table 3.4 (Nagamachi 2001). A development may adapt one or any combination of these six types or develop new types. For example, using virtual reality technology and simulation

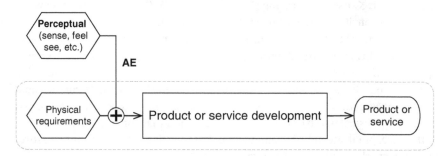

**Figure 3.15** AE on customer's perceptual expectations.

**Table 3.4** Types of AE methods.

| Type | Methods | Function |
| --- | --- | --- |
| 1 | Strategy and decision | Translation of consumer's feelings and image to design elements |
| 2 | Software | Computer system with AE databases |
| 3 | Hybrid design | Interaction of user's impressions and design specifications |
| 4 | Modeling | Mathematical modeling |
| 5 | Virtualization | Computer integration of virtual reality technology |
| 6 | Collaborative design | Using software and databases over the internet |

*Source*: Nagamachi, M. (2001). Workshop 2 on Kansei Engineering, Proceedings of International Conference on affective Human Factors Design, June 27–29, 2001, Singapore.

can visualize the operations and functions of a product or service to address the affective features and their effects (Häfner et al. 2012).

In his later book, Dr. Nagamachi introduced the five basic process steps of applying AE (Nagamachi and Lokman 2011). The steps are:

1. Identify customers and determine how to handle their emotional expectations.
2. Determine product concept, by surveying and studying customer lifestyle.
3. Break down the product concept to sub-concepts (to product characteristics).
4. Deploy physical design characteristics, such as weight and color.
5. Translate characteristics to technical specifications.

In Step 3, development professionals translate a product concept from a customer study to physical product characteristics. This is an engineering practice called Type I by Dr. Nagamachi. In translating customer expectations to physical design characteristics, one may build a mathematical model between inputs and outputs (physical characteristics). One can also study other types of customer desires or feelings, and translate them into new product features, which is Type II of AE. Type III is unique to the specific modeling method used, which may be one of those listed in Table 3.4, or a new method altogether.

With AE introduced, the overall development process for a new product or service remains unchanged. However, the considerations and design efforts based on customer feelings are added and integrated into every step, particularly the early ones, of a development process. Figure 3.16 shows AE's role in product or service development.

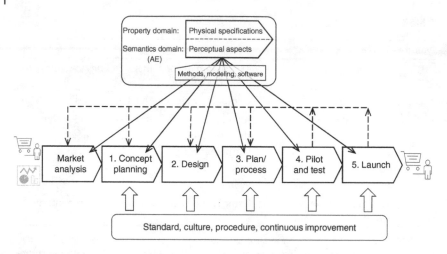

**Figure 3.16** Translation process of customer's demands and perceptions.

### 3.4.2 Discussion of AE

**Considerations to Applications**

An important step of AE is to apply semantics, to define and select Kansei – (i.e. affective – words, to describe customer perceptions of a product or service (Dahlgaard et al. 2008). The sources for deriving Kansei words include market analysis, domain experts, experienced users, trade magazines, previous affective studies, and websites. For example, one may select a number of Kansei words for a flower or bouquet including beautiful, clean, pure, fresh, elegant, rich, bright, cheerful, genial, sweet, etc. (Kanamori et al. 2005). Another example is a battery-powered hand tool. The Kansei words for a tool may include value, ease of use, comfort, low vibration, appealing styling, long battery life, and bright color. If one develops a new product that bears an old brand name (or to reintroduce/rebirth a product), some words like nostalgia, reminiscent, or sentimental could be considered as Kansei words. Note good Kansei words should solicit emotional imagery when they are read, as opposed to simply describing physical facts. While cold water may define the physical state of that water, "crisp" or "refreshing" may provide a more somatic image, and as such may be considered a better Kansei word for describing that water.

With Kansei words, the remaining task is to link and translate them into design requirements and generate innovative solutions. This connection may be made using different approaches. For example, one may consider category or cluster analysis (Sato 2017), regression analysis (Dolgun and Köksal 2018), fuzzy theory (Liu et al. 2019), or rough set theory (Akgül et al. 2020).

**Table 3.5** Design evaluation example using Kansei words

| Kansei word | Value | Comfort | Vibration | Styling | Battery life | Color | ... | Weighted score |
|---|---|---|---|---|---|---|---|---|
| Weight ($W_i$) design | 15.2 | 9.5 | 5.7 | 10.4 | 5.2 | 5.8 | ... | |
| 1 | 5 | 3 | 3 | 4 | 3 | 4 | | 202.0 |
| 2 | 4 | 4 | 4 | 5 | 4 | 5 | | 223.4 |
| 3 | 5 | 4 | 3 | 5 | 3 | 5 | | 227.7 |
| 4 | 3 | 5 | 5 | 3 | 4 | 4 | | 196.8 |

One simple method is to evaluate design options using about 10 key Kansei words where each has an associated weight. The weights are based on customers' expectations and perceptions. After all weights ($w_i$, $i = 1$ to n, n is the number of Kansei words.) are determined, they are rescaled (or normalized) to ($W_i$) so the total of weight is 100. The calculation of $W_i$ is:

$$W_i = \frac{w_i}{w_1 + w_2 + \dots + w_n}$$

From the calculation, $W_i$ is likely a decimal number. So are the weighted scores. The weighted scores in Table 3.5 are calculated based on the data shown in the table.

The rating for each Kansei word is often on a scale of 1–5; level 5 is the best at meeting customers' desires. If a Kansei aspect is quantitatively measurable, say a battery life, it can be converted into a 1–5 scale. From the weighted scores, one can ascertain the top contributors for customer satisfaction, and create a priority list as a design guide.

**Characteristics of AE**

AE is predictive in nature for product or service development. It is a unique approach to help innovative development and technology push (discussed in Chapter 2) line up with customer demands, and differentiate a new product and service from the competition. Adopting AE also extends the developmental aims and scope of work, which implies that a product and service is made for both physical/biological needs and mental/psychological wants.

Because of the focus of AE, it is complementary to the Kano model for AQ and QFD in identifying customer requirements, and converting them into producer's specifications. AE techniques can help identify and prioritize what QFD refers to as "subjective" and "unconscious" quality attributes. For example, the value of a product or service can be a subjective and intangible attribute. In the same way, appreciation of the style of a product can depend on the individual.

As a relatively new approach, AE is actively an early stage of research and development. One study found some characteristics and/or limitations of existing AE applications (Bongard-Blanchy et al. 2013):

- Most evaluated in the aspects of forms and colors.
- Project-specific, and difficult to apply universally in other contexts.
- Time-consuming to implement.
- Evaluated mostly based on static images of finished products, instead of real objects.
- Limited during the conception stage, lack of the interaction between consumers and development objects.
- Mainly done on tangible factors of predefined product components.
- Does not consider the changes that take place over time.

The nature of customer perception itself challenges the application of AE, because the perception varies from customer to customer, changes for a given customer over time, and is affected by multiple other factors. These complex situations and factors should be addressed in new studies and applications of AE.

Another challenging question is how to balance the various identified needs and wants during development. Any analysis, like Table 3.5, may be only for a concept evaluation on an affective side, while a comprehensive evaluation should combine both physical and perceptual factors. Another question could be how to weigh the factors of needs vs. wants. These considerations and limitations could lead to additional research efforts.

### 3.4.3 Applications of AE

The principles of AE make sense, and can apply to all customer-related industries. One focus of applied study is on the evaluation and analysis of customer perception, because it is subjective in nature. What are a few practices of AE that can be useful references for new applications.

Vieira et al. (2017) reported that AE was used for the design of in-vehicle rubber keypads. The silicone rubber keypads are commonly used as push-switches in in-vehicle multimedia products. The main physical parameters in a keypad design are actuation force, contact force, stroke, and snap ratio. Applying AE, they collected 116 words via e-mail for the subjective evaluation of keypad interface in a semantic space. They then categorized the words into five groups and selected higher-level Kansei words to represent the concept of the groups. The

Kansei words of the groups were global evaluation (unpleasant/pleasant), depth (short/long), feedback (with/without sharp click), force (smooth/hard; not very/very strong), and stability (fragile/robust; loose/stiff). Seven pairs of words in parentheses were then used for the evaluation to represent the semantic space. Twenty-eight volunteers participated in the evaluation. After data analysis, strong relations were found between subjective perceptions and physical properties and suggested the guidelines for the construction and evaluation of new rubber keypads (Vieira et al. 2017).

Service quality planning and improvement are typical applications of AE. A study on luxury hotel services found that the service attributes deserved more attention, regarding their impacts on customer emotion (Hartono 2012). The author integrated the SERVQUAL model, Kano model, and QFD to capture customer emotional needs in the hotel service study. Then the author built a mathematical model to represent the relationships between Kansei words and service attributes. The author considered five dimensions of human interaction service quality: tangibility, reliability, responsiveness, assurance, and empathy in their survey for 181 hotel guests. From the study results, the author proposed a guideline for service managers on which service attributes were sensitive to customer delight, and with a priority for continuous improvement.

The more a product's or service's market share is directly related to subjective appeal (literature, architecture, fashion, art, etc.), the more AE can play a crucial role in development, bridging the gap between people and technologies. One case study on fashion design supported this notion (Wang et al. 2018). The interactive clothing study combined artistic design perspectives with information-sensing engineering methods and Kansei evaluation method. Wang and his colleagues established the correlation between properties of clothing and their evolutionary path. Their Kansei evaluation of survey for non-interactive and interactive was composed of two steps. The first step used a semantic differential method to analyze the effects of the interactive clothing on the psychological responses of wearers. Survey questions were designed in eight levels: arousal, excitement, joy, fun, composure, laziness, sadness, and anger. In the second step, they followed up with a comparative experiment of multi-category smart clothing evaluation. They found that their prototypes correlated well with human emotional and expressive patterns (Wang et al. 2018).

AE, as a relatively new and distinctive approach, can be more effective in the development of a new product or service when integrating with other technologies. Cooperatively applying AE with other methods and technologies has proven to indicate many potential benefits for an organization.

# Summary

## Goal: Design for Customer

1 Customer satisfaction is not only a customer's reflection on a product or service but also a study process.
2 A customer study can be on promotional, understanding, procedure, performance, and perception gaps, to reveal how to achieve and improve customer satisfaction.
3 A common method of customer study is surveys. Others include interviews, focus group studies, mystery shopping, related data analysis, etc.
4 There are several national or regional inter-industry measure indexes, e.g. ACSI.
5 Product or service development follows a systematic realization process, to translate customer demands into the functions and features of a product or service.
6 Characteristics of a product, service, or process can be categorized into the three levels of critical, special, and ordinary, based on their relative significance and other factors.

## Quality Category to Customer

7 MBQ is about the attributes of a product or service meeting basic customer expectations. As being expected, MBQ improvement may slightly increase customer satisfaction.
8 AQ is about the attributes of a product or service that can pleasantly surprise and delight customers. AQ can be a sales booster and provide a competitive advantage.
9 PQ (or one-dimensional quality) shows a quality attribute performance proportional to customer satisfaction.
10 The Kano model, mainly by applying MBQ, AQ, and PQ, distinguishes the different influences of quality attributes relative to customer satisfaction.
11 One of the key tasks in applying a Kano model is to establish the connections between the quality dimensions and quality attributes of a product or service.

## QFD

12 QFD is a proactive planning approach that builds a connection among customer expectations, product/service characteristics, and market competition for a product or service.

**13** A QFD structure, or house of quality, has four segments: customer expectations, product/service characteristics, market and rival evaluation, and technical specifications.

**14** The QFD approach emphasizes the relationship between the four segments, qualitatively and/or quantitatively.

**15** For a complex product or service, a QFD may apply multiple times, at different stages of development. For multiple QFDs, the relationship among them should be established.

**16** As a common practice, QFD is applied jointly with other approaches.

**AE**

**17** AE addresses customers' feelings and perceptions, and transforms them into executable design parameters and actions.

**18** The original AE consists of six types of methods; new methods can be introduced to AE applications.

**19** The AE approach should be integrated into the development steps of a product or service.

**20** A common practice of AE is to identify and select Kansei words, and then address them in development.

## Exercises

### Review Questions

**1** Discuss the roles of customer study and benchmarking study for new product or service development.

**2** Discuss the translation process of customers' needs into product or service development with an example.

**3** Find an example and discuss the three levels of characteristics for a product or service.

**4** Provide your definition, and example, of MBQ.

**5** Provide your definition, and example, of AQ.

**6** Find examples to fit the four levels of quality and customer satisfaction (Figure 3.6).

**7** Explain the MBQ and AQ curves and their shapes in a Kano model.

**8** Provide an example of PQ.

**9** Provide an example of irrelevant quality.

**10** If a product fails or becomes popular, which quality attribute (MBQ, AQ, or PQ) is most likely the main reason? Why?

**11** Describe the overall structure and elements of a QFD.

12 Explain how to evaluate the interrelationship in a QFD.
13 Discuss the concept of multi-level QFDs.
14 Explain the concept and principle of AE.
15 List a few input variables for AE.
16 One states that AE and Kano model are complementary to each other to address customer expectations. Do you agree? Why, or why not?
17 Discuss integrating AE into a conventional development process.
18 Create five Kansei words for a product or service and discuss their influences on customers.
19 Discuss how to determine the weights of Kansei words.
20 Discuss a challenge in using AE.

**Mini-project Topics**

1 Select an industry and study its national customer satisfaction index.
2 Review the relationship between a product (or service) characteristic and process (or procedure) characteristic.
3 Locate a research paper on Kano modeling from library databases, or https://scholar.google.com, and review the application of the Kano model.
4 Develop a simple QFD, based on a product or service you are familiar with.
5 Locate a research paper on AE from library databases, or https://scholar.google.com, and review the application of AE.
6 Study the scenario of multi-level QFD, and discuss benefits and potential issues of the application.
7 Locate a research paper on the combined application of QFD and another method from library databases, or https://scholar.google.com, and review the application.
8 Select a product or service case and apply Kano modeling (Figure 3.8).
9 If you plan to develop a new service, describe how to use Dr. Nagamachi's five steps of applying AE.
10 Locate a research paper on integrating AE with another approach from library databases, or https://scholar.google.com, and review the application.

# References

Akgül, E., Özmen, M., Sinanoğlu, C. et al. (2020). Rough Kansei mining model for Market-oriented product design, *Mathematical Problems in Engineering*, Vol. 2020, Article ID 6267031. 10.1155/2020/6267031.

Alsharqawi, M., Zayed, T., and Dabous, S.A. (2018). Integrated condition rating and forecasting method for bridge decks using visual inspection and ground penetrating radar, *Automation in Construction*, Vol. 89, pp. 135–145. 10.1016/j. autcon.2018.01.016.

Altuntas, S. and Kansu, S. (2019). An innovative and integrated approach based on SERVQUAL, QFD and FMEA for service quality improvement: A case study, *Kybernetes*, 10.1108/K-04-2019-0269.

Anderson, E.W. (1998). Customer Satisfaction and Word of Mouth, *Journal of Service Research*, Vol. 1, Iss. 1, pp. 5–17. 10.1177/109467059800100102.

ASCI. (2020). Benchmarks by industry, The American customer satisfaction index, https://www.theacsi.org/acsi-benchmarks/benchmarks-by-industry, accessed in December 2020.

Avikal, S., Singh, R., and Rashmi, R. (2020). QFD and fuzzy Kano model based approach for classification of aesthetic attributes of SUV car profile, *Journal of Intelligent Manufacturing*, Vol. 31, pp. 271–284. 10.1007/s10845-018-1444-5.

Balvers, R., Gaski, J., and McDonald, B. (2016). Financial disclosure and customer satisfactions do companies talking the talk actually walk the walk?, *Journal of Business Ethics*, Vol. 139, Iss. 1, pp. 29–45.

Bongard-Blanchy, K., Bouchard, C., and Aoussat, A. (2013). Limits of Kansei – Kansei unlimited, *International Journal of Affective Engineering*, Vol. 12, Iss. 2, pp. 145–153. 10.5057/ijae.12.145.

Buttigieg, S.C., Dey, P.K., and Cassar, M.R. (2016). Combined quality function deployment and logical framework analysis to improve quality of emergency care in Malta, *International Journal of Health Care Quality Assurance*, Vol. 29, Iss. 2, pp. 123–140. 10.1108/IJHCQA-04-2014-0040.

Camgöz-Akdağ, H., Tarım, M., Lonial, S. et al. (2013). QFD application using SERVQUAL for private hospitals: A case study, *Leadership in Health Services*, Vol. 26, Iss. 3, pp. 175–183. 10.1108/LHS-02-2013-0007.

Chen, C.C. (2010). Application of quality function deployment in the semiconductor industry: A case study, *Computers & Industrial Engineering*, Vol. 58, Iss. 4, pp. 672–679. 10.1016/j.cie.2010.01.011.

Chen, K.J., Yeh, T., Pai, F. et al. (2018). Integrating refined Kano model and QFD for service quality improvement in healthy Fast-food chain restaurants, *International Journal of Environmental Research and Public Health*, Vol. 15, Iss. 7, pp. 1310. 10.3390/ijerph15071310.

Chen, M.C., Hsu, C.L. et al. (2019). Service quality and customer satisfaction in pharmaceutical logistics: An analysis based on Kano model and importance-satisfaction model, *International Journal of Environmental Research and Public Health*, Vol. 16, Iss. 21, pp. 4091. 10.3390/ijerph16214091.

Chen, S., Pai, F., and Yeh, T. (2020). Using the Importance–satisfaction model and service quality performance matrix to improve long-term care service quality in Taiwan, *Applied Sciences*, Vol. 10, Iss. 1, pp. 17. 10.3390/app10010085.

Choi, S.B. and Kim, J.M. (2018). A comparative analysis of electronic service quality in the online open market and social commerce: The case of Korean young adults, *Service Business*, Vol. 12, Iss. 2, pp. 403–433. 10.1007/s11628-017-0352-7.

Dahlgaard, J.J., Schütte, S., Ayas, E. et al. (2008). Kansei/affective engineering design: A methodology for profound affection and attractive quality creation, *The TQM Journal*, Vol. 20, Iss. 4, pp. 299–311. 10.1108/17542730810881294.

Dehe, B. and Bamford, D. (2017). Quality function deployment and operational design decisions–a healthcare infrastructure development case study, *Production Planning & Control*, Vol. 28, Iss. 14, pp. 1177–1192. 10.1080/09537287.2017.1350767.

Dolgun, L.E. and Köksal, G. (2018). Effective use of quality function deployment and Kansei engineering for product planning with sensory customer requirements: A plain yogurt case, *Quality Engineering*, Vol. 30, Iss. 4, pp. 569–582. 10.1080/08982112.2017.1366511.

Gangurde, S.R. and Patil, S.S. (2018). Benchmark product features using the Kano-QFD approach: A case study, *Benchmarking: An International Journal*, Vol. 25, Iss. 2, pp. 450–470. 10.1108/BIJ-08-2016-0131.

Ginting, R., Hidayati, J., and Siregar, I. (2018). Integrating Kano's model into quality function deployment for product design: A comprehensive review, IOP Conference series: Materials science and engineering, Vol. 319, *The 4th Asia Pacific Conference on Manufacturing Systems and the 3rd International Manufacturing Engineering Conference*, December 7–8, 2017, Yogyakarta, Indonesia. 10.1088/1757-899X/319/1/012043.

Gonzalez, M.E. (2019). Improving customer satisfaction of a healthcare facility: Reading the customers' needs, *Benchmarking: An International Journal*, Vol.26, Iss. 3, pp. 854–870. 10.1108/BIJ-01-2017-0007.

Gonzalez, M.E., Quesada, G., Martinez, J.L. et al. (2019). Global education: Using lean tools to explore new opportunities, *Journal of International Education in Business*, Vol.14, Iss. 1, pp. 37-58. 10.1108/JIEB-11-2018-0052

Grapentine, T. (2015). Why the Kano model wears no clothes, Quirks Marketing Research Media, Article ID: 20150407, https://www.quirks.com/articles/why-the-kano-model-wears-no-clothes, accessed in August 2020.

Gremyr, I. and Raharjo, H. (2013). Quality function deployment in healthcare: A literature review and case study, *International Journal of Health Care Quality Assurance*, Vol. 26, Iss. 2, pp. 135–146. 10.1108/09526861311297343.

Grigoroudis, E. and Siskos, Y. (2009). *Customer Satisfaction Evaluation: Methods for Measuring and Implementing Service Quality*. Berlin: Springer.

Häfner, P., Ommeln, M., Katicic, J. et al. (2012). Immersive Kansei engineering – A new method and its potentials. https://www.imi.kit.edu/downloads/ommeln-2012a.pdf, accessed in October 2020.

Hartono, M. (2012). Incorporating service quality tools into Kansei engineering in services: A case study of Indonesian tourists, *Procedia Economics and Finance*, Vol. 4, pp. 201–212. 10.1016/S2212-5671(12)00335-8.

Horton, G. and Goers, J. (2019). A revised Kano model and its application in product feature discovery. Preprint, https://www.researchgate.net/publication/332304132_A_Revised_Kano_Model_and_its_Application_in_Product_Feature_Discovery, accessed in May 2020.

ISO. (2021). *ISO 16355-1 Application of Statistical and Related Methods to New Technology and Product Development Process – Part 1: General Principles and Perspectives of Quality Function Deployment (QFD)*. Geneva: International Organization for Standardization.

Jennings, J. (2019). Carbon Fiber QFD, QUAL 552 Quality Planning (CRN 13393), Eastern Michigan University, September 13, 2019.

Jia, W., Liu, Z., Lin, Z. et al. (2016). Quantification for the importance degree of engineering characteristics with a multi-level hierarchical structure in QFD, *International Journal of Production Research*, Vol. 54, Iss. 6, pp. 1627–1649. 10.1080/00207543.2015.1041574.

Kanamori, T., Imai, Y., and Takeno, J. (2005). Extraction of 430,000 association words and phrases from Internet web sites, *Proceedings of the 2005 IEEE International Conference on Mechatronics and Automations*, pp. 1929–1934, Niagara Falls, Canada, July 20–August 1, 2005. 10.1109/ICMA.2005.1626857

Kano, N., Seraku, N., Takahashi, F. et al. (1984). Attractive quality and must-be quality, Hinshitsu (Quality), *The Journal of the Japanese Society for Quality Control*, Vol. 14, Iss. 2, pp. 39–48.

Koprivica, C., Bešić, S.M., and Ristić, O. (2019). QFD method application in the process of hotel service quality improvement, *Hotel and Tourism Management*, Vol. 7, Iss. 2, pp. 57–66. 10.5937/menhottur1902057M.

Kurniawan, R. (2017). Passenger's perspective toward airport service quality – Case study at Soekarno-Hatta International Airport, *Journal of the Civil Engineering Forum*, Vol. 3, Iss. 1, 10.22146/jcef.26547.

Leber, M., Ivanišević, A., and Borocki, J. (2018). Fostering alliances with customers for the sustainable product creation, *Sustainability*, Vol. 10, Iss. 9, pp. 3204. 10.3390/su10093204.

Li, Y., Liu, S., Xu, L. et al. (2015). Decision target adjustment quality function deployment network with an uncertain Multi-level programming model for a complex product, *Journal of Grey System*, Vol. 27, Iss. 3, pp. 132–150.

Liu, S.F., Hsu, Y., and Tsai, H. (2019). Development of a new cultural design process using Kansei engineering and fuzzy techniques: A case study in Mazu crown design, *International Journal of Clothing Science and Technology*, Vol. 31, Iss. 5, pp. 663–684. 10.1108/IJCST-12-2017-0183.

Lo, S.M., Shen, H., and Chen, J. (2017). An integrated approach to project management using the Kano model and QFD: An empirical case study, *Total Quality Management & Business Excellence*, Vol. 28, Iss. 13–14, pp. 1584–1608. 10.1080/14783363.2016.1151780

Löfgren, M., Witell, L., and Gustafsson, A. (2011). Theory of attractive quality and life cycles of quality attributes, *The TQM Journal*, Vol. 23, Iss. 2, pp. 235–246. 10.1108/17542731111110267.

Lorenzi, C. and Ferreira, J. (2017). Failure mapping using FMEA and A3 in engineering to order product development: A case study in the industrial automation sector, *International Journal of Quality & Reliability Management*, Vol. 35, Iss. 7, pp. 1399–1422. 10.1108/IJQRM-10-2016-0179.

Ma, H., Chu, X., Xue, D. et al. (2019). Identification of to-be-improved components for redesign of complex products and systems based on fuzzy QFD and FMEA, *Journal of Intelligent Manufacturing*, Vol. 30, Iss. 2, pp. 623–639. 10.1007/s10845-016-1269-z.

MacDonald, E., Backsell, M., Gonzalez, R. et al. (2006). The Kano method's imperfections, and implications in product decision theory, *Proceedings of the 2006 International Design Research Symposium* (pp. 1–12). Seoul, Korea, November 10–11.

Mansur, A., Farah, A.N., and Cahyo, W.N. (2019). Integration of servqual, Kano model, and QFD to design improvement on public service system, *IOP Conference Series: Materials Science and Engineering, Volume 598, Annual Conference on Industrial and System Engineering (ACISE) 2019* April 23–24, 2019, Semarang, Central Java, Indonesia. 10.1088/1757-899X/598/1/012101.

Mayne, E. (2007). Chrysler wary of too much success with Swivel 'n Go, WardsAuto, May 9, 2007. https://www.wardsauto.com/news-analysis/chrysler-wary-too-much-success-swivel-n-go, accessed in August 2020.

Mikulić, J. and Prebežac, D. (2011). A critical review of techniques for classifying quality attributes in the Kano model, *Managing Service Quality: An International Journal*, Vol. 21, Iss. 1, pp. 46–66. 10.1108/09604521111100243.

Nagamachi, M. (2001). Workshop 2 on Kansei Engineering, *Proceedings of International Conference on affective Human Factors Design*, June 27–29, 2001, Singapore.

Nagamachi, M. and Lokman, A.M. (2011). *Innovations of Kansei Engineering*. Boca Raton, FL: CRC Press Taylor & Francis Group.

NIST. (n.d.). Baldrige performance excellence program, The National Institute of Standards and Technology, US Department of Commerce, https://www.nist.gov/baldrige, accessed in September 2020.

Poel, I. (2007). Methodological problems in QFD and directions for future development, *Research in Engineering Design*, Vol. 18, pp. 21–36. 10.1007/s00163-007-0029-7.

Power, J.D. (2020 a). New-vehicle quality mainly dependent on Trouble-free technology, J.D. Power Finds, Press Release, June 24, 2020. https://www.jdpower.com/business/press-releases/2020-initial-quality-study-iqs, accessed in August 2020.

Power, J.D. (2020 b). Some new vehicle technologies risk failing while others become "must have," J.D. Power Finds, Press Release, August 19, 2020. https://www.jdpower.com/business/press-releases/2020-us-tech-experience-index-txi-study#:~:text=Volvo%20ranks%20highest%20overall%20with,)%20and%20Genesis%20(559), accessed in August 2020.

Pun, K.F., Chin, K.S., and Lau, H. (2000). A QFD/hoshin approach for service quality deployment: A case study, *Managing Service Quality: An International Journal*, Vol. 10, Iss. 3, pp. 156–170. 10.1108/09604520010336687.

Sajid, A. (2020). 70 Best Startups You Need to Watch Out for in 2020, https://www.cloudways.com/blog/best-startups-watch-out/#ai, accessed in August 2020.

Sato, H. (2017). Kansei evaluation for the front grill design of automobile, *Transactions of Japan Society of Kansei Engineering, 2017*, Vol. 16, Iss. 1, pp. 51–60. 10.5057/jjske.TJSKE-D-16-00040.

Schneider, B., Macey, W.H., Lee, W.C. et al. (2009). Organizational service climate drivers of the American customer satisfaction index (ACSI) and financial and market performance, *Journal of Service Research*, Vol. 12, Iss. 1, pp. 3–14. 10.1177/1094670509336743.

Seddon, J. (2005). *Freedom from Command & Control: Rethinking Management for Lean Service*. New York: Productivity Press.

Serugga, J. (2020). A utilitarian decision – making approach for front end design – A systematic literature review, *Buildings*, Vol. 10, Iss. 34, 10.3390/buildings10020034.

Shahin, A., Iraj, E.B., and Shahrestani, H.V. (2018). Developing the C-shaped QFD 3D Matrix for service applications with a case study in banking services, *International Journal of Quality & Reliability Management*, Vol. 35, Iss. 1, pp. 109–125. 10.1108/IJQRM-02-2016-0018.

Shahin, A., Pourhamidi, M., Antony, J. et al. (2013). Typology of Kano models: A critical review of literature and proposition of a revised model, *International Journal of Quality and Reliability Management*, Vol. 30, Iss. 3, pp. 341–358. 10.1108/02656711311299863.

Shaker, F., Shahin, A., and Jahanyan, S. (2019). Developing a two-phase QFD for improving FMEA: An integrative approach, *International Journal of Quality & Reliability Management*, Vol. 36, Iss. 8, pp. 1454–1474. 10.1108/IJQRM-07-2018-0195.

Sherbini, A., Aziz, Y.A., Sidin, S.M. et al. (2017). Differences between service quality and customer satisfaction: Implications from tourism industry, *International Journal of Applied Business and Economic Research*, Vol. 15, pp. 343–360.

Sivasamy, K., Arumugam, C., Devadasan, S.R. et al. (2016). *Advanced models of quality function deployment: A literature review, Quality & Quantity*, Vol. 50, pp. 1399–1414. 10.1007/s11135-015-0212-2.

Sun, K.A. and Kim, D.Y. (2013). Does customer satisfaction increase firm performance? An application of American Customer Satisfaction Index (ACSI), *International Journal of Hospitality Management*, Vol. 35, pp. 68–77.

Tang, H. (2019). A new method of bottleneck analysis for manufacturing systems, *Manufacturing Letters*, Vol. 19, pp. 21–24. 10.1016/j.mfglet.2019.01.003.

Vieira, J., Osório, J., Mouta, S. et al. (2017). Kansei engineering as a tool for the design of in-vehicle rubber keypads, *Applied Ergonomics*, Vol. 61, pp. 1–11. 10.1016/j.apergo.2016.12.019.

Vinodh, S., Manjunatheshwara, K.J., Sundaram, S.K. et al. (2017). Application of fuzzy quality function deployment for sustainable design of consumer electronics products: A case study, *Clean Technologies and Environmental Policy*, Vol. 19, Iss. 4, pp. 1021–1030. 10.1007/s10098-016-1296-7.

Wang, W., Nagai, Y., Fang, Y. et al. (2018). Interactive technology embedded in fashion emotional design – Case study on interactive clothing for couples, *International Journal of Clothing Science and Technology*, Vol. 30, Iss. 3, pp. 302–319. 10.1108/IJCST-09-2017-0152.

Xu, K., Chen, Y.V., Zhang, L. et al. (2019). Improving design software based on fuzzy Kano model: A case study of virtual reality interior design software, *13th International Conference of the European Academy of Design*, Dundee, April 10–12, 2019. 10.1080/14606925.2019.1594923.

Yadav, H.C., Jain, R., Singh, A.R. et al. (2017). Kano integrated robust design approach for aesthetical product design: A case study of a car profile, *Journal of Intelligent Manufacturing*, London, Vol. 28, Iss. 7, pp. 1709–1727. 10.1007/s10845-016-1202-5.

Zaim, S., Sevkli, M., Hatice Camgöz-Akdağ, H. et al. (2014). Use of ANP weighted crisp and fuzzy QFD for product development, *Expert Systems with Applications*, Vol. 41, Iss. 9, pp. 4464–4474. 10.1016/j.eswa.2014.01.008.

# 4

# Quality Assurance by Design

## 4.1 Design Review Process

### 4.1.1 Introduction to Design Review

**Design Process and Review**

Product and service design are creative, multi-step processes. In such a process, there are many requirements, steps, and tasks associated with the specific product or service to be designed, as they might pertain to a given discipline. A general design process for a complex product or service is composed of five phases, as shown in Figure 4.1. In the design phase, there are three major tasks: system design, detail design, and verification tests.

An initial design for a new product or service is rarely perfect, as any initial design is subject to the knowledge, skills, and efforts of designers, as well as resource constraints. Therefore, it is necessary to conduct what is called a design review in general. This general review includes checks on the intended functions, brainstorming activities, and trials for solutions of unforeseen problems, with the purpose of making adjustments to reach design goals.

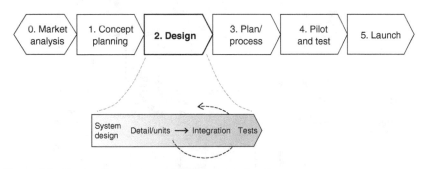

**Figure 4.1** Design process in product or service development.

*Quality Planning and Assurance: Principles, Approaches, and Methods for Product and Service Development*, First Edition. Herman Tang.

The aim of a design review is to evaluate the design of a product; its major components, or a function thereof, against given design intents and requirements. Based on review and predefined requirements, one can gauge the design status, and determine the extent of a need to refine a design (or even redesign). Thus, a design review is a quality assurance process throughout the course of a design phase in product or service development. In addition, a review process used in design can also apply to later phases of development, including manufacturing planning, prototyping, release, and launch. Given this perspective, a design review is an integral part of both product and service development in their entirety.

The purpose, focus, and format of design reviews can vary from one project to another, even within the same field. One may need to conduct various design reviews to support design success and effectiveness, summarized in Figure 4.2 and discussed in the next subsection.

### Types of Design Review

Depending on the goal relative to a design review, e.g. functionality, cost assessment, new technology application, or problem solving, there are six types of design review. For a simple product or service, some types of review may be combined or skipped, while the process flow would stay the same.

1. <u>State-of-the-art review:</u> To benchmark the competitors' products. A state-of-the-art review is often conducted by a marketing department. From this review, a "wish list" of features is created as a reference to what a new product or service should look like. Hence, a state-of-the-art review should be conducted before a requirement review.

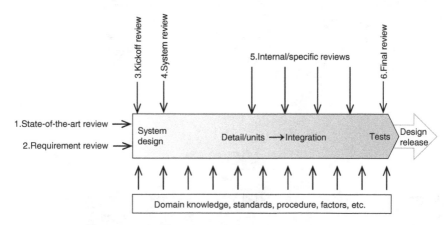

**Figure 4.2** Various reviews in design process.

2. <u>Requirement review:</u> To identify customer expectations, propose design require-ments, and assess feasibility. The inputs include the outputs from customer sur-veys, state-of-the-art reviews, product trials, manufacturing capacity, warranty information, and regulation requirements. Affective engineering (AE) approaches (discussed in Chapter 3) should be implemented in this review.

3. <u>Design kickoff review:</u> To start the review on concept selection, main param-eters, technical adequacy, and general compliance with the predefined requirements. Typically, a kickoff review is a milestone of a design project.

4. <u>System design review:</u> To finalize overall requirements and main features, and allocate them to subsystems or parts. A system design review is sche-matic, and offers a bird's-eye view on the entire design project, as directives for detail designs are instituted from the kickoff.

5. <u>Internal/critical design reviews:</u> To evaluate key elements, normally focus-ing on specific subjects, e.g. a performance, progress, or prototyping result. Such reviews may be conducted multiple times, even for the same purposes. These reviews frequently concern the results of special tryouts and tests.

6. <u>Final verification review:</u> To assess and verify a design prior to submission for approval. A final review should be conducted with, or by, the executives responsible for the project. In a final review, a full-scale test is often needed. A successful final review of a product design means a project completion and subsequent design release.

### Characteristics of Design Review

Design review is called "review," but is actually development, evaluation, and creative work. For example, a requirement review may bring out the concerns of must-be quality (MBQ) and attractive quality (AQ), and investigate how to develop attractive and affective quality features. With such features, a new product or service could be appealing in terms of customer satisfaction and be competitive in the market.

Teamwork is central to design review. To achieve design goals, design reviews can be conducted by different ways and means. For example, a design review can be informal and frequent within design teams, while a final review is nor-mally planned and executed by a cross-functional team. Working with seasoned professionals and learning from other domain practices can be one of the best ways to conduct a design review. Depending on the review purpose, a design review can have different participants.

One challenge to consider, given the variance in how a review is done, is the maintaining of records for these efforts. Review documentation, including agenda, meeting minutes, issue tracking, and follow-up plan, is critical. Depending on the type of review, the corresponding documents, such as project timeline, specifications, test results, drawings, layouts, and budget, should be prepared before a review meeting.

Another challenge is time. The plan of a design (and other) phases is often developed based on time available. A short time frame may lead to some shortcuts with less than ideal reviews taking place. The efficiency and path of a design review in the given time may give the best chances of a design success, which also relies on the managerial skills and experience of those involved with the review.

Design reviews have been widely implemented in product design, including physical products and software. As discussed, a service and its quality differ from products in many aspects, but the overall development procedure for a product and service can be very similar. Besides, physical products and services often are packaged together. Design reviews then apply to the integrative development of both products and services.

### 4.1.2  Design Review Based on Failure Mode

#### Principle of Design Review Based on Failure Mode

A special design review process, developed and used by Toyota, is called design review based on failure mode (DRBFM, also known as the Mizenboushi method). The basic idea of DRBFM is to address the potential problems in a new design, or modification to an existing proven design.

DRBFM is a three-step process: 1) good design, 2) good discussion, and 3) good dissection to achieve a reliable new or modified design. This process also is sometimes called $GD^3$, shown in Figure 4.3.

1. <u>Design:</u> Proposing a new design or major design change, a design team starts by looking at the existing designs and standards and implementing "best practices."
   - The aim of a new design or major modification is to have the same or better outcome as the existing ones.

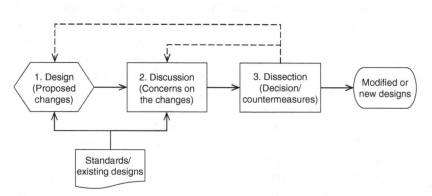

**Figure 4.3**  Process flow of DRBFM.

- A good design may require specifying the design conditions of a reliable product and use these proven conditions for a modified or new design as much as possible. A design team identifies what makes the design robust, and how to adapt it for a new design.
- A major change or new design often comes with a risk of low reliability, or high likelihood of quality issues. Simultaneous changes in different areas of the same design can have higher risks because the changes may influence each other. Thus, incremental, small, and non-complex (not simultaneous, not in multiple places) changes are preferred.

2. Discussion: A discussion is on the potential impacts of proposed design changes or new designs, as they concern the customers.
   - As the core of DRBFM, the purpose of a good discussion is to identify design weaknesses and potential issues.
   - Discussion focuses may be on changes on quality, processes, costs, and suppliers associated with potential issues identified.
   - Many techniques and methods, such as risk analysis and assessment (discussed in Chapter 2), can be used in the discussion step.

3. Dissection: A dissection review is an in-depth discussion and review, focusing on countermeasures to potential issues and conducting tests if applicable.
   - Problem identification and determination of the countermeasures for the potential problems are engaged.
   - Evaluation tests, such as a reliability test, may be necessary to a dissection review. For example, a given dissection may specifically address test results on the signs of wear or degradation.

**Discussion of DRBFM**

The DRBFM method is preventive in nature because it focuses on the future risks and potential issues introduced by changing an existing design or creating a new one. DRBFM should be conducted during the design stage, instead of in a later testing stage, as the latter can lead to last-minute extra design changes and/or design release delay. In principle and process, DRBFM is very similar to design failure modes and effects analysis (FMEA), which is discussed in Chapter 5.

Toyota requests a DRBFM or design FMEA for all parts. A design change should follow the failure mode identified through DRBFM development. Otherwise, the potential failure mode would have to be addressed on a later process FMEA to have inspection and detection capabilities in manufacturing processes. However, that may increase the quality cost discussed in Chapter 1. General Motors also requires a DRBFM for every fit/form/function change.

To track the potential issues identified in a DRBFM, a worksheet can be used, as shown in Table 4.1. After the occurrence frequency (O), significance to customers (S), and difficulty to detect (D) are rated, the risk priority numbers (RPNs) of all items are calculated to prioritize work.

**Table 4.1** DRBFM worksheet format

| Product: | | Owner: | | | | | Date: | | | DR #: | | | |
|---|---|---|---|---|---|---|---|---|---|---|---|---|---|
| 1. Part number | 2. Part name | 3. Function | 4. Design change | 5. Potential issues | 6. Possible cause | 7. Other factors | 8. Occurrence frequency (O) | 9. Significance to customers (S) | 10. Difficulty to detect (D) | 11. RPN = O × S × D | 12. Recommendation | 13. Responsibility and timing | 14. Conclusion and follow-up |
| (Contents) | | | | | | | | | | | | | |

DRBFM makes sense for the development of products and services in many industrial applications, as these products and services evolve from existing ones. For example, automakers introduce new vehicle models with limited redesigns from the current ones. In such gradual product changes, DRBFM is an effective tool to sustain new product quality.

DRBFM applications are still mainly in the automotive industry; SAE has a recommended practice SAE J2886 (SAE 2013). Application examples include vehicle design (Wright 2011) and electronic control system (Yamaguchi et al. 2019). As a preventive analysis tool, DRBFM, at least its principle and process, can be used for other industries.

The invention of a product or service can introduce new risks and potential problems. One can have extra tasks and burdens from an innovative design based on DRBFM. As a tool to study potential issues proactively, for an improved or modified design, a DRBFM should be coordinated with not hinder or slow an innovation process. In addition, a creative design can be more important than having a perfect design in many cases. A brand-new product, such as Tesla's electric vehicles, or new services, such as Uber and Lyft ridesharing taxi services, relied on few proven designs or references. Their business successes are mainly from their innovations, rather than design perfection. For example, Tesla received an initial quality score of 250 PP100 (problems per 100 vehicles) for 2020, significantly worse than most automakers (Power 2020). Innovation should be encouraged without the intimidation of risks and additional workloads and safeguarded by proactive procedures.

### 4.1.3 Design Review Applications

In early development stages, a design review can be conducted to address current situations, initial designs, and challenges. For example, the Divertor Tokamak Test Facility (DTT), a high field superconducting toroidal device, was a major construction approved by the Italian government. The DTT carries plasma current up to 5.5

MA in pulses to develop a reliable solution for power and particle exhaust in a magnetic confinement fusion reactor (Martin 2019). The design review of the DTT mainly regarded engineering aspects, and led to a vertical symmetric tokamak after thorough reviews. In those reviews, the professionals started with the definition of a vertical symmetric basic machine, changed some crucial dimensions with consideration of budget constraints, and planned additional power coupled with the high performance plasma with increased flexibility. They also considered the radial build as a critical aspect of the DTT construction. The design reviews additionally made substantial changes on the magnet system, the vacuum vessel, the neutron shield, and the thermal shield, with a significant improvement in efficiency (Albanese et al. 2019). From this example, one may learn that a design review can significantly change and improve an original design based on various technical, build, and financial factors.

Design reviews can be beneficial to analyze current situations and challenges as well. For instance, in Northern Ireland, a design review was conducted at a national level, to explore the evolution of the practices and some challenges faced (Mackel and Kinneir 2019). In a later development phase, a design review can be used as a process to gain stakeholders' feedback and improve design. One study was on the use of design review meetings to address the maintainability design in the post-occupancy phase of a building, including owners, occupants, and facility operators (Liu et al. 2018). The review process should include a reviewing cycle, information handling, and decision-making.

As another example, one design review evaluated the principal models of cable driven exoskeleton, focusing on the upper limbs. The review was on different requirements, ranging from the anatomical to the transmission system, noting the approaches used to tackle the complexities of each joint and transmission system (Sanjuan et al. 2020). The review provided a reference for the early stages of the design of exoskeletons. An earlier design review was conducted on a baby stroller. This reviewed the baby stroller designs (patents) during 1980 to 2014 and found several feature evolutions, such as foldability, ergonomic designs, and electronic applications (Sehat and Nirmal 2017). The authors of the review consequently proposed new functions and features focused on comfort and convenience.

## 4.2 Design Verification and Validation

### 4.2.1 Prototype Processes

#### Objectives of Prototyping
In the development of a product, particularly a large or complex one, there is an embedded process called prototyping. A prototype is one or a series of preliminary creations or working models, which are built as a tangible or visible

presentation for demonstration, evaluation, and/or validation purposes. In this sense, most prototypes are bonded with design reviews, that themselves often are based on prototypes, either physically or virtually.

A prototype, as a premature model, can differ from the final product in various ways, in terms of functionality, material, appearance, etc. The types of prototypes include:

- Concept: To illustrate the ideas, often in a format of virtual or animation and in a reduced scale.
- Rapid: To show a future product for its functionality, features, and/or feasibility, as a product not made of actual intended materials.
- Mockup: To provide a full size but nonfunctional product, often for internal and detail reviews.
- Test: To be used for various purposes to test e.g. user experience and specific functions, sometimes close to its actual product. Figure 4.4 shows a prototype of a stationary wireless charging system for an electrical vehicle battery (Galigekere et al. 2018).

In addition to a physical form, a prototype can be in various formats using virtual reality, simulation or animation technologies. A prototype may be built in wireframe, clay, 3D printing, etc. For other types of products, such as manufacturing process, software, technology, or service, they can have additional formats. Vehicle prototype parts are often built using so-called "soft tools," which are simplified, temporary, and non-production-intended tooling.

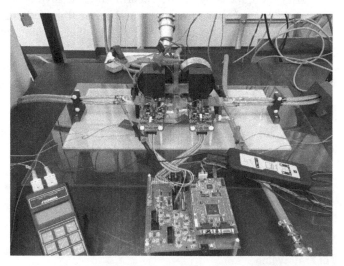

**Figure 4.4** Prototype of wireless battery charging system. *Source*: Galigekere, V.P., Pries, J., Onar, O. et al. (2018). Design and implementation of an optimized 100 kW stationary wireless charging system for EV battery recharging. https://www.osti.gov/servlets/purl/1495980.

In testing processes, a prototype may fail in one way or another, which serves one of testing's purposes: to reveal design problems and issues. The design team will find solutions and revise their design. However, not all problems can be found through prototypes and tests.

The concept and process of prototyping can be used for any type of product. As a collaborative and circular process, prototyping takes considerable efforts and helps a new product become successful, win new business, and avoid various types of waste.

### Types of Prototype Process

As an early and preliminary model, a prototype build normally applies to a product's development to test and try out the concept, functionality, and/or build process of a product. A prototype test can be conducted either physically or virtually, and a prototype itself can be full-sized or a scale model, made of temporary materials.

A prototype build plan should specify its description and objectives, including the materials, functions, and performance tests, and corresponding timing. Many organizations and government agencies have standardized procedures for prototyping. For example, the US Department of Energy supports commercial prototype building models (DOE n.d.). If a US company develops a new medical device, its development should follow the US FDA's process and focus on safety. The device development process of the FDA has five steps, where Step 2 is about prototyping (FDA 2018a):

1. Device discovery and concept. This first step normally begins in a laboratory.
2. Preclinical research – prototype. The second step is to do laboratory and animal testing on safety.
3. Pathway to approval. Following a successful Step 2 prototype, a new device is tested on people for safety and effectiveness.
4. FDA review. FDA teams thoroughly examine all the submitted data related to the device and decide on approval.
5. FDA post-market safety monitoring. Once a device is available for public use, FDA monitors its safety.

A service prototype differs from a product prototype, as a service is normally intangible. Service development follows a similar process, where a service prototyping sometimes is called service testing. In service prototyping phases, the service development team may simulate, evaluate, and review the planned service for various aspects, such as its scope, requirements, procedure, timing, customer interaction, and potential issues. Therefore, the principles of prototyping are generic and applicable for both product and service development.

In terms of format, service prototyping varies. Common formats of service prototypes. For example, a video prototype is used to represent the scenario to show a customer's interaction with a service in real settings. Using role-playing prototyping may help understand the design intents and options, and verify solutions. Similar to role-playing, service walkthrough simulates the processes and systems of a service in a full-scale, real setting, in a later phase of development. When actual customers are involved, a test process is called live prototyping, which can be used for final validation purposes. Discussed in earlier chapters, a product and service can be combined as a bundle. In such cases, prototyping can be conducted with multiple steps with different methods.

## 4.2.2 Processes of Verification and Validation

### Concepts of Verification and Validation

The US government's definition of verification is "confirmation by examination and provision of objective evidence that specified requirements have been fulfilled" (CFR 2012). The verification approach has been widely used in everyone's daily and professional life. For example, a computer login often needs a second means of verification, such as using a cell phone or recognizing pictures to confirm.

Design verification is a process to examine and provide evidence that the design output of a product or service meets the intended specifications. A design verification may use various physical and/or computer simulation tests on specific attributes, such as functional requirements for useful life, robustness under predefined conditions, and failure avoidance. Overall, design verification, as an internal process, is to examine whether the product has been designed right.

The US government's definition of design validation is "establishing by objective evidences that device specifications conform to user needs and intended use(s)" (CFR 2012). Design validation is the process of evaluating the new product or service developed at the end of the development cycle. The purpose of validation is to determine if the designed product or service satisfies the specifications regarding customers' needs. A validation process often involves acceptance and suitability with external customers. As a whole, a design validation is to ask whether the right product has been designed.

Design verification and validation are conducted through analyses, demonstrations, inspections, and/or tests, physically or virtually. Design verification and validation normally do not include the effects of manufacturing-induced issues and variation, which should be addressed separately.

**Table 4.2** Differences between software verification and validation

| Verification | Validation |
|---|---|
| To check if developing the software correctly to the specs | To check if developing the correct software to customers |
| Low-level, detailed focused exercise | High-level exercise |
| Done by a quality assurance team | Done by a testing team |
| To conduct static practice (no executing code) | To conduct dynamic testing (with executing code) |
| Manual checks via reviews, meetings, inspections, walk-throughs, etc. | Various types of computer-based testing, such as black-box (functional), gray-box, white-box (structure) |

A difference between verification and validation is that validation is related to customer's expectations, while verification is related only to the designed product's performance. Besides, verification and validation are conducted at different times and scales. Validations are carried out after verifications. For software engineering, verification and validation can be checked with various questions, and by then asking those questions in different ways (Table 4.2). Sometimes, the terms verification and validation are used interchangeably.

**Characteristics of Verification and Validation**
A common characteristic of design verification and validation is well-defined specifications and requirements, which serve as the foundation for verification and validation. Like any type of development project, verification and validation require a good plan, execution, and follow-up.

One system, the V-shape model, has been widely used in product development. Its left wing encompasses development tasks and activities. The right wing encompasses the quality assurance activities corresponding to those development activities. The objectives of the right wing are verification and validation. A key characteristic of a V-shape model is the connections between the two wings. Figure 4.5 overviews the verification and validation in a typical development process. In development phases, verifications often have different objectives and focuses, e.g. functionality and performance. In addition, different verifications have unique procedures and methods, e.g. product reliability (Zhou and Li 2009).

Figure 4.6 shows the process flow of verification and validation in a design phase. In other development phases, there may also be verification and validation processes and tasks. For example, a purpose of the launch of a product or

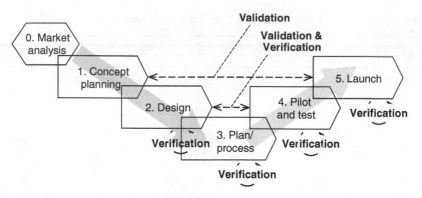

**Figure 4.5** Overview of verification and validation in development.

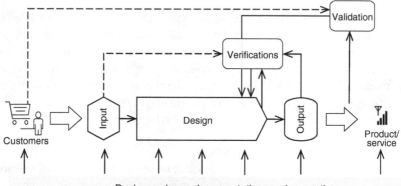

Design reviews, documentation, and correction

**Figure 4.6** Verification and validation in design.

service is to validate whether it meets the requirements determined in the early concept planning. The prototyping and tests are carried out to validate if the design is satisfying customer expectations.

Note that verification and validation is confirmative (to approve design performances) rather than exploratory (to find out and deal with "what if" situations). Naturally, a formal verification should be planned at the end of a design phase, while subsystem verification tasks can be conducted during that design phase.

Sometimes, verification and validation are all called testing. For example, software development has five types of testing: unit, integration, system, alpha, and beta testing. Alpha testing is an internal acceptance test before releasing a software product to end users. Beta testing is a test performed by real users in a real environment before shipping the software to customers. Both alpha and beta testing are validation tests on customer expectations.

In a given industry, there may be industry-specific standards for validation practices. For example, the FDA has the guideline FDA-2008-D-0559 for manufacturing human and animal drug and biological products (FDA 2018b). For software engineering, there is an approach called Independent Software Verification and Validation process to guide validation (IEEE 2017). For greenhouse gas, there is the standard ISO 14064 Specification with guidance (ISO 2018).

### 4.2.3   Discussion of Verification and Validation

#### Considerations in Verification and Validation

As an integral process of development, verification and validation may be in different forms and/or with different considerations, while serving similar purposes. Here are a few examples of verification and validation applications. For example, design verification tests were conducted for an LED driver (Başol et al. 2019). In this verification, the authors considered the driver's worst-case scenarios and implemented the worst cases into tests.

Verification may be conducted in the form of an audit. A study assessed the impact of the entire audit process based on the pre-audit data, audit data, and post-audit data (Giganti et al. 2019). They found a significant discrepancy between datasets, and suggested that improving data quality following the audit may affect study results.

In preparing, conducting, and completing a design verification, professionals may need to fill out a formal record, or form. One example would be a Design Verification Plan and Report (DVP&R), found in the automotive industry. The purpose of using a DVP&R is to document the planned tests and results for a component or subassembly. This report is usually in a spreadsheet that outlines all necessary tests, the main conditions of the tests, methods used, specifications, acceptance criteria, sample sizes, and responsible personnel (Figure 4.7). A DVP&R is an element of a Production Part Approval Process (PPAP) package, which will be discussed in Chapter 6.

For a complex part and product, design verification itself can be a small project, requiring a design test flow, testing facility capacity, required equipment, and workforce. If a test result does not meet specifications or objectives, a corrective action plan should be included in the report for following up.

A design validation is typically utilized at a system level. In addition to design inputs and requirements, direct or simulated customers are also involved in a validation process. For example, a design of a human hand exoskeleton was tested on actual people, in addition to computer simulations (Hansen et al.

Administrative information
(Part name, number, supplier info, date, track number, supervisor, etc.)

| Test | Descript. | Method/Spec | Sample | Due Date | Assigned | | Results | Conclusion |
|------|-----------|-------------|--------|----------|----------|--|---------|------------|
| 1. | Vibration | Instrument, process, requirements, criteria | 10 | 11/24/20 | Responsible person | | Test report | Pass and approved |
| | | | | | | | | |
| | | | | | | | | |
| | No blanket statement is acceptable. | | | | | | | |

Authorized signature, date, etc.

**Figure 4.7** Sample DVP&R form.

2018). Another example affirmed this notion, wherein certified diabetes educators used the medical device design of a glucose monitor to conduct a comparative ease-of-use analysis to confirm usability (North et al. 2019). In another validation work, researchers used a real-life wind speed profile to test a laboratory-scale wind turbine emulator (Ajewole et al. 2017).

**Advance of Verification and Validation**

Many practitioners and researchers try to find more effective ways to conduct design verification. Here are a few examples. New approaches for design verification and validation keep coming out.

- Two authors modeled the uncertainty for a product design verification by using a theory of belief function also known as the Dempster–Shafer theory (Kukulies and Schmitt 2018).
- A study suggested design verification through virtual prototyping techniques based on systems engineering (Mejía-Gutiérrez and Carvajal-Arango 2017).
- Another study used a reasoning process (Mehrpouyan et al. 2016).
- Some authors proposed a planning model and considered the priorities of the failure modes based on failure rates, detectability, and consequences (Mobin et al. 2019).

A common assessment during verification and validation is to provide a rating for the critical characteristics of a product or service. The rating criteria can be based on completion relative to risk, in three levels of indicator: green, yellow, and red (see Table 4.3) for a quick identification in a checklist.

**Table 4.3** Assessment criteria example of verification and validation.

| Status | Color (symbol) | Criteria description |
|---|---|---|
| Full complete/ no risk | Green (√) | Deliverables on track and meeting objectives and requirements |
| Most complete/ moderate risk | Yellow (?) | Deliverables not meeting target dates and/or requirements; recovery work plan approved and in place |
| Partial complete/ high risk | Red (×) | Deliverables not meeting target dates and/or requirements; no recovery work plan available |

Virtual technologies, e.g. virtual reality and augmented reality, have been used for verification and validation. For example, a control technology called virtual commissioning has been used in manufacturing system development (Lechler et al. 2019, Alkureidi 2020). In service, virtual reality becomes a tool for study and validation, e.g. healthcare (Loannou et al. 2020). There is still a lot of work to be done developing virtual technology to a state on par with actual processes and tasks. Verification and validation in a virtual environment is a burgeoning field; an active research and development subject with great opportunities.

## 4.3 Concurrent Engineering

### 4.3.1 Principle of Concurrent Engineering

#### Concept of Concurrent Engineering

Concurrent engineering (CE), sometimes referred to as simultaneous engineering, is not a branch of engineering but a systematic approach for new product and service development. It is called "engineering" probably because it was initially developed for engineering design to integrate with its related manufacturing processes simultaneously. CE as a principle and approach has been widely implemented in various disciplines, such as project management and business development.

The concept of CE is to incorporate and conduct some relevant tasks concurrently with cross-functional teams, which was traditionally done sequentially. Figure 4.8 depicts a comparison of sequential development and CE-based development. In the traditional way, for example, a manufacturing process is to be planned after the product design is completed. Applying CE, the process planning should start during design rather than waiting until design completion.

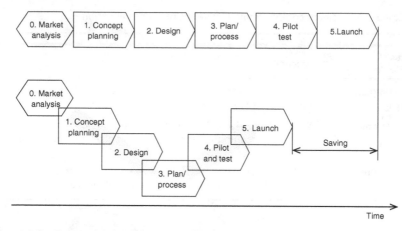

**Figure 4.8** Sequential development vs. CE-based development.

The CE principle can also apply to detailed designs, with multidisciplinary teams working at the same time. For a product with software and hardware, some design tasks may be conducted in parallel, regardless of whether software dictates most of the hardware features or vice versa. A simultaneous, collaborative process can make design work effective, and aid immensely in resolving development issues.

The core of CE is the practice of organic integration of multiple related development tasks, with the intention of observed concurrency and synchronization of the integrated tasks. Without a meaningful, well-prepared integration plan, a CE implementation would be just a multi-tasking work style, which probably creates unnecessary conflicts and issues. Collaboration, in terms of people, information, and knowledge, remains a major challenge for CE implementations (Stjepandić et al. 2015).

An investigation was once conducted on the concept and definitions of CE. The authors offered insights on the similar concepts of CE and collaborative engineering (CoE) (Putnik and Putnik 2019). They considered that CE was more on the overlapped execution of processes, but lacked "concurrency." When emphasizing the collaboration among cross-functional teams, CoE might be a better word to describe the nature of such efforts. Their viewpoint may lead to in-depth discussion of the characteristics of CE.

### Process of CE

Applying the CE principle, many development tasks and reviews are conducted at the same time, as shown in Figure 4.8. When multiple teams are working on different aspects of development simultaneously, the design team normally plays a technical central role. Figure 4.9 shows an example of product design

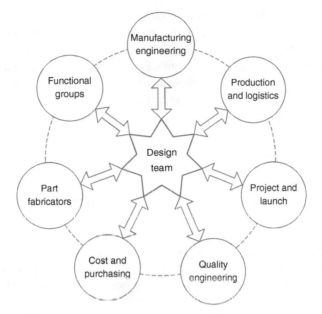

**Figure 4.9** Cross-disciplinary teamwork for CE implementation in product design.

engineering working with other departments for a complex product. In such a cross-disciplinary working environment, the hub of a CE wheel is design engineering. The arrows (spokes of the wheel) are the input and output communication interfaces of the design team with other groups. Upper management coordinates and administers the work of multiple teams.

One common CE process involves presenting a preliminary design solution to a cross-functional team for inputs. The cross-functional team reviews the design from their unique perspectives. The design team considers these inputs, arriving at a revised solution to satisfy as manyparties as possible. Such a process often needs multiple iterations. Toyota practices a different method, wherein the design team develops and communicates the *sets* of preliminary solutions to the cross-functional teams (Sobek et al. 1999). When narrowing their respective sets of solutions, the teams continue to refine and communicate the set to a final, optimized design. Set-based CE or set-based design has been an active research topic.

Implementing CE can make the development process more efficient, lead to shorter time to market, and better the entire system rather than locally optimizing subsystems. As an additional benefit, CE encourages multidisciplinary collaboration, which can bring to the surface many issues that can occur in later development phases. For example, the collaboration between product design engineers and manufacturing process engineers can bring about the identification and solutions for the manufacturability issues because of uneconomic designs. The tangible improvements from CE applications include high quality and low costs, as the efforts are made proactively to prevent or reduce the defects. On the management side, CE may help foster an improved collaborative culture within the company.

### 4.3.2 Considerations to CE

#### Applications of CE

Various industries have implemented CE as a common practice. A few examples are cited here as references, and readers may search for specific ones in an interesting discipline. For example, one paper summarized the applications of systems CE to aerospace products in the past 20 years (Loureiro et al. 2018). The authors found that the system solutions contained both product and service elements, and analysis was performed simultaneously for product and service elements.

A study by Fischer et al. (2017) showed a conceptual data model for the CE process at the German Aerospace Center. The model was in a tree-like hierarchical decomposition (like the event tree analysis – ETA – discussed in Chapter 2) to break down complexity and organize the information into logical blocks such as system, subsystem, and equipment. The high-level model was used for discussions with the different disciplinary users at different locations, which allowed them to understand what information they could store and use in the database. The authors implemented the data model in software, to share the information, and this allowed design teams to analyze "on the fly."

A survey investigation was conducted on the impacts of total quality management (TQM) and CE on performance, based on over 200 respondents of manufacturing firms (Deshpande 2019). The study results implied that companies with TQM and CE implementations enhanced their manufacturing performance, and that TQM and CE practices complemented each other well. It also suggested that other quality management elements, such as supplier quality management, quality planning, and quality process monitoring, could be implemented for their impacts on CE and manufacturing performance.

CE can also apply to integrating engineering and business functions. For example, Arnette and Brewer (2017) showed that a high CE intensity improved procurement activity and product performance. Practitioners and researchers continue to study CE, to make its applications more broad and successful.

**Concerns of CE Implementation**

CE implementation has its own challenges, and its implementation achievements are mixed. First, a business system has to provide time commitment, schedule, and structure for cross-disciplinary work. Technical communication collectively among the professionals in different fields requires a large bandwidth, and it can be difficult to reach a consensus collectively among the different disciplines. Therefore, the compromise on major issues should be based on a predefined procedure and mutual understanding. Adopting CE principles may experience corporate culture related issues. One study listed the principal areas of concerns: controlling the direction of projects, conflicting goals of personnel, short-term requirements, blame culture, commitment to new procedures, and management of the engineering department (Filson and Lewis 2000).

Not every development task nor every moment in development requires cross-functional teamwork, or to be done in parallel with another task. Some tasks must be sequential, or part of a larger pipeline. Management needs to plan and administer tasking and time frame for a proper CE work environment and implementation.

Another key challenge is that cooperative work requires the coordination of knowledge resources. CE not only involves the accomplishment of multiple tasks in parallel but also works to ensure effective interaction among functional groups. A shared computer environment is essential for CE implementation. For example, issues between the design and manufacturing process need to be discussed in a shared computer design environment, which should be friendly and efficient for all participating parties. Such an environment often means a significant investment, in computer technology and training up front. Additionally, cooperative work is often even more important when teams work remotely.

CE related tasks often add more workload and complexity in development stages. Based on the inputs from other teams, the design staff may have to do an analysis to respond, and then carry out modifications, which takes additional time. For the entire development, CE efforts can benefit products or services because the issues resolved could be costly if they happen later. This notion agrees with the opinions of experienced professionals such as Andeson (2020). Nevertheless, such upfront efforts might be unfavorable to some design professionals.

**CE in Service**

Originally developed in product design, CE has established its generality for application in many disciplines. There is a dedicated scholarly journal *Concurrent Engineering: Research and Applications* (ISSN: 1063-293X), launched in 1993, for publishing the research and applications arising from parallelism of product life cycle development.

Service development, such as in healthcare, housekeeping, tourism, nursing, and teaching, differs from product design in several ways, as discussed in Chapter 1. For example, a service execution has different requirements regarding the operational processes and equipment of a manufacturing production. Accordingly, it is necessary to build different models for service development based on the CE principle. Figure 4.10 shows an example.

In this CE implementation process for service development, concept design and system design have a noteworthy overlap. Detailed or departmental designs are comprehensive, with significant coordination among all departments. The overall process of CE may be the same for service development, while the specific steps depend on the detailed contents of a service design.

For service industries and service-oriented business operations, there have been many CE applications. For instance, a study on the supply chain for hotels showed that CE might enhance efficiency of the supply chain, reduce product costs, increase profits, and lead to the gaining a competitive advantage (Union et al. 2020).

Abbasi and Nilsson's (2016) study on the themes and challenges in developing environmentally sustainable logistical activities proposed a holistic and integrative model that was built on a three-dimensional CE framework. Their three-dimensional CE model focused on sustainability, with pillars emphasizing the role and competence of logistics service providers in contributing to more sustainable development of products, processes, and supply chains.

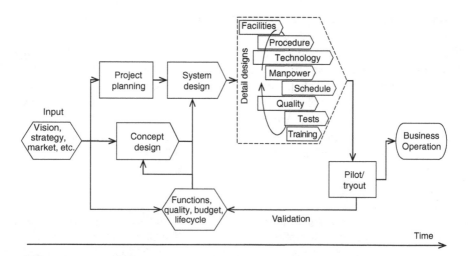

**Figure 4.10** CE implementation process flow in service development.

# 4.4    Variation Considerations

## 4.4.1    Recognition of Variation

### Understanding Variation

Variation can be viewed as a change or difference in an activity, process, or result, in terms of condition or amount. Variation always exists, anywhere and forever. Just like Dr. Deming (1994) said, variation is life and life is variation. For example, there are always differences in products and services, operational performances, environmental factors, and customer expectations.

Because of variation, the quality of any product or service cannot be 100% perfect, and always has some types of defects. The defect or failure rate can be very low for some operations, e.g. air transportation. Through understanding variance and its impact on customers, one can better plan tasks, in both design and execution, to reduce variance.

There are many factors affecting a variation. If the influencing factors are denoted by $x_i$ ($i = 1, 2, 3, ...$) and the resulting variation $y$, then the relationship between the factors and variation can be generally presented as.

$$y - f(x_i), \ i - 1,2,3,...$$

where $f$ stands for function, $f$ itself often is a theoretical unknown. Even if the relationship function between a variation and its factors can be found and a corresponding model established, the function normally is nonlinear and complicated. Therefore, understanding and reducing variation is based on a statistical analysis of actual data.

Commonly practiced in new product and service launch, execution, and routine operations, variation reduction is a major focus of continuous improvement of product and service quality. A goal of variation reduction usually targets improving quality, efficiencies, and reducing costs.

There are systematic studies on variation and its impacts, e.g. in fields of biology (Hallgrimsson and Hal 2011), design (Weber and Duarte 2017), teaching (Huang and Li 2017), and linguistics (Eckert 2018). However, considering and studying variation in the quality planning and assurance of new development has not been widely implemented, and the research body on this subject contains limited publications. The subject of proactive design-in-quality in products and services requires more research into its methods and applications.

### Sources of Variation

One can categorize the sources of variation in two general types: common causes and special causes. Either of these types can be of the known or unknown variety. Common causes are inherent in a system, sometimes called "natural or inherited causes," bringing random variance into a product, service, or process.

Poor product or service design, inappropriate process design, inadequate machinery, and environmental conditions are typical common causes for the variation of a product or service. It is often assumed that the data of common-cause variation follows a normal distribution.

The design and planning of a product or service determine the types and amounts of common causes in variation. Hence, the common causes of variation should be addressed in the early development phases (Figure 4.11). The practice of proactively addressing and minimizing common-cause variation is also called quality by design. Usually, such natural causes cannot be removed by continuous improvement in executions, only by redesigning the product, process, or service. Therefore, it is difficult, if not impossible, to reduce common-cause variation in operations. However, it is possible to compensate for common-cause variation.

Special causes are mostly due to operational conditions and parameters, including reduced quality of incoming materials and parts, out-of-spec environmental conditions, lack of training for operators, and inadequate equipment maintenance. Thus, special causes also are called "assignable causes" or "correctable problems." In most cases, special causes result in a large and unexpected variation for a process. From this point, addressing special causes to reduce variation is an effective way to improve quality and customer satisfaction (Mclaughlin 1996), which is not in the scope of this book.

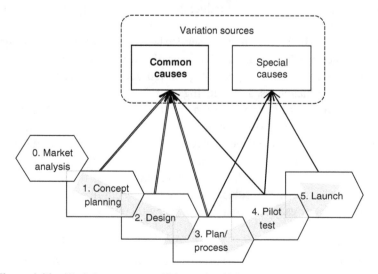

**Figure 4.11** Work focuses on variation reduction.

Special causes of variation can affect the prototyping, launch, and operations of a product or service. Sometimes, special causes may also be related to the lack of robustness in a design. Therefore, in later phases of development, e.g. process planning and pilot testing, both common and special causes should be addressed as opportunities to reduce resultant variation.

### 4.4.2 Target Setting with Variation

#### Variation of Product and Service

Target setting is about defining the expected achievements in terms of the near future. As discussed in Chapter 1, target setting can be done surrounded by many knowns and unknowns, as these can be presented by a certain variation around a numerical target. With the understanding of variation, one can view a quality target in two basic parameters: one is the mean value ($\mu$) and the other is the standard deviation ($\sigma$) with a normal distribution assumption.

An engine of a passenger car may be selected as an example for discussion. A market study shows that the lifespan of customer expectation for an engine is 150,000 miles (about 241,407 km). Because the lifespan of the engine has variation, the question is: What should be the design lifespan to make 95% of engines meet customer expectation? Here is a simplified analysis considering variation.

First, the engine design management team should know the existing manufacturing quality capacity for the current engines. Without introducing new design features and manufacturing technologies, the quality of the engines is likely to stay at the same level. For simplicity, the $\sigma$ of the engine lifespan is assumed to be 10,000 miles. If the design target is $\mu = 150,000$ miles, then unsatisfied customers are about 50% as shown in Figure 4.12 (a). To ensure that 95% engines meet the 150,000-mile lifespan target, the actual design target (or the average lifespan) should be 166,450 miles, as illustrated in Figure 4.12 (b).

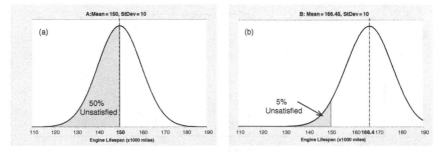

**Figure 4.12** Example of engine design target setting.

From this simple case, one can know that variation is an important factor for quality target setting.

Other factors should be considered, such as adopting new technologies, reducing variation by continuous improvement, and/or having certain budget constraints. Then, the final target could be significantly different because the data distribution curve may take a different shape.

**Variation of Customer Expectation**

It is equally important to consider the variation of customer expectations in a target-setting process. Customer expectations can vary in many ways, such as age, gender, geographic region, background, and overtime. Unlike the variation of a product and service, the variation of customer expectations is dynamic and likely uncontrollable by an organization.

The variation of customer expectations is attainable through market research. When considering variation of customer expectations, the relationship between the customer expectation and engine lifespan is shown in Figure 4.13. In this figure, the larger the distance "d" between the expectation and target is, the higher customer satisfaction likely is.

Now, there are two datasets: 1) engine lifespan and 2) customer expectation. They are assumed to be independent and following normal distributions. In the figure, $f(x_c)$ and $f(x_e)$ present the probability function of the customer and engine, respectively. Then, one can predict the probability that some engines do not satisfy the customers.

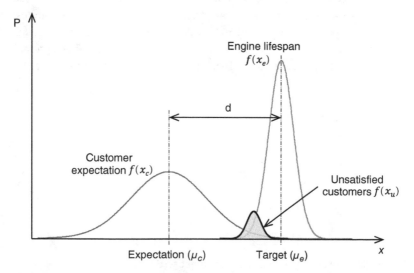

**Figure 4.13** Target-setting modeling with consideration of variations.

This concept introduces the variation in customer expectation in a target-setting process, in addition to a conventional approach only based on the variation of product reliability. The distributions of both datasets are often known. If the probability (P) of unsatisfied customers is determined, then the design target ("d" in the figure) can be calculated. For example, if the customer expectation is 150k miles and unsatisfied customers should be less than 3%, then the targeted engine lifespan can be calculated.

Considering the variation of customer expectations is an effort to change an important unknown factor to a calculated factor. Thus, this is better than the conventional consideration of customer expectations as a general "safety factor" in target setting. This viewpoint of variation is in line with reliability engineering, which deals with the estimation, prevention, and management of the "lifetime" of product uncertainty and risks of failure.

### 4.4.3 Propagation of Variation

#### Element Quality and Assembly Quality
A product, service, or manufacturing process is a system comprising many elements. Each element has its own characteristics and quality attributes, which may propagate and eventually show up in the final product or service. It is important to know not only what the quality characteristics and attributes of elements are but also how they contribute to overall quality for customers.

As discussed in Chapter 1, service quality normally has five main dimensions (Table 1.3). Each dimension directly affects overall service quality, as shown in Figure 4.14, in the format of a fishbone diagram (more discussion on this tool in

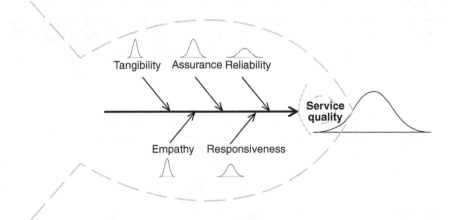

**Figure 4.14** Service quality and its elements.

Chapter 8). The remaining question becomes the relationship between each dimension and overall service quality. For example, how much does quality tangibility contribute to overall quality? Different from physical products, such a relationship is mostly qualitative, and pending more studies to become quantitative.

Individual quality characteristics and their influences have been studied. For example, a serial manufacturing process comprises multiple operations. Each operation has a quality issue rate $p$ ($p \geq 0$) as shown in Figure 4.15. At the end of a process, the probability of the final quality without quality issues is:

$$Final\ quality\ rate = (1-p_1)(1-p_2)(1-p_3)\ldots(1-p_n) = \prod_1^n (1-p_i)$$

where $n$ is the number of operations.

For a simple case, all operations are assumed to have the same rate $p$ (called a homogeneous process). The final assembly rate *without* a quality issue would be $(1-p)^n$. If $p = 1.0\%$ and $n = 10$, then the probability of the good final assembly is $(1-p)^n = 90.44\%$. This simple case shows how the quality of each element affects the total quality of a product or service. Real-world cases can be more complex. For example, a complex system comprises many elements in serial, parallel, and mixed settings, has different individual rates, and has different types of issues. The variations of individual components can also be enlarged/lessened during transmission toward the final product.

**Complexity of Quality in Systems**
The quality of components may contribute to the final product, depending on a few factors, such as the specific quality attributes of a component and the architecture of a final product. An example of the dimensional quality of a door assembly in relation to the quality of a complete car is shown in Figure 4.16.

Here, the dimensional variation of a door assembly affects the size of the door gap for a car. The door gap is not only an esthetic issue, but also affects

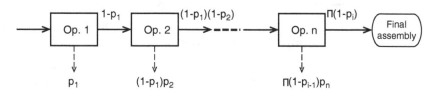

**Figure 4.15** Individual quality and final quality in serial operations.

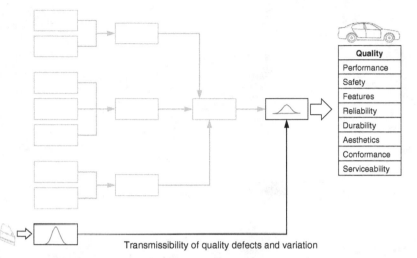

Transmissibility of quality defects and variation

**Figure 4.16** Relationship between component quality and assembly quality.

the performance of door-closing effort, wind noise, and water leak prevention. Adapting the concept of the transfer function in engineering, the quality relationship between a door and car in an assembly system can be presented (*f* standing for function):

$$Q_{car} = f(q_{door})$$

The relationship can be modeled as a reflection, which may be in a linear mode as shown in Figure 4.17. A good design should make the slope of the reflection line near to zero; meaning no component quality variation influences the quality of the final product.

Such a relationship can be complicated for a complex product or service. For example, a reflection line could be nonlinear, shown as a curved dash-line in the figure. Understanding the relationships between individual components and their final assembly helps design the *appropriate* levels of quality for components, to avoid making them unnecessarily "too good" if some components do not really matter to the final assembly.

Another concern is about possible interactions with other components. A minor quality issue of a component alone may be acceptable or not visible on the final product. However, a minor issue can become unacceptable because of the quality of neighboring or mating components. For example, if a door opening is in a borderline dimensional quality in the same direction with the

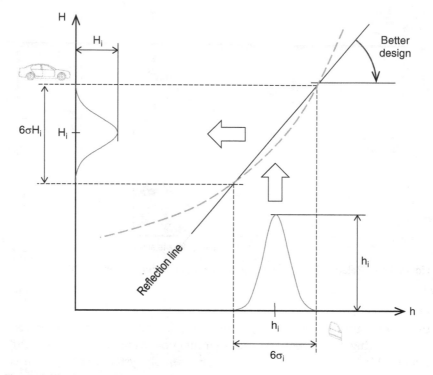

**Figure 4.17** Reflection modeling of variation transmissibility.

variation of a door, the door gap variation may become intolerable. Some factors are unknown, e.g. residual stress of the materials, which can worsen the dimensional variation of a door and door opening. Considering multiple factors, a general expression of a final product (e.g. car) may be represented:

$$Q_{car} = f\left(q_1, \ldots, q_{door}, q_{opening}, \ldots, q_n\right)$$

More comprehensive and specific studies are needed on this challenging subject of quality and variation transmission, particularly for major components. For example, an engine has significant contributions to the complete product. Such studies can be a valuable reference to product design and manufacturing operations.

### 4.4.4 Quality and Variation

#### Variation and Specifications
In quality planning, the requirements and technical specifications of a new product or service should be defined and designed with consideration of

variation. The three key points of variation related to quality planning have been reviewed already:

1. Variation always exists, including the variation of customer expectations.
2. Variation matters in quality target setting, including specification design.
3. The variation of components transmits to the final product.

Quality specifications, as quantitative requirement statements, in product design are based on customer expectations, product functionality, and manufacturing capability. For individual components, additional consideration is given to their contributions and interactions with other components to the quality specifications of an entire product. From these points, it is arguable that to define and determine quality specifications can be more of an art than a science in most cases.

Conventional thinking surrounding quality specifications is that they are on the borderline between customer satisfaction and dissatisfaction, or called good quality and poor quality. A product is deemed to be good quality when its feature and variation are inside the quality specifications. In addition, from a product producer or service provider standpoint, quality specifications are closely associated with costs; the tighter the specifications, the higher the costs. To ensure customer satisfaction, a higher level of quality specifications is preferred, when they are feasible and justifiable, based on resources.

However, it is a myth that if a feature of a product does not meet its specification limits, then customers can immediately see the quality issue and be dissatisfied. Such a conventional concept, sometimes called a "goalpost" philosophy, is illustrated in Figure 4.18 (a). In the figure, "t" is a target value; LSL and USL are the lower and upper specification limits, respectively. In reality,

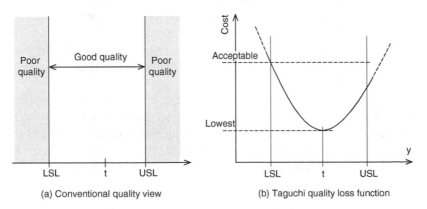

(a) Conventional quality view          (b) Taguchi quality loss function

**Figure 4.18**   Conventional view of quality and Taguchi loss function.

customers probably do not feel that way. Besides, the level of customer dis-satisfaction is not the same when a quality is at a borderline level or entirely unacceptable.

**Taguchi Loss Function**

A better perspective regarding variation and customer satisfaction is the Taguchi quality loss function. Its concept states that a drop in quality does not suddenly change customer satisfaction. Customers have an expectation target, and their satisfaction level is lowered when the quality of a product or service deviates from their value. The loss may be presented by a parabolic function in concept:

$$L = k \times (y - t)^2$$

where, $t$ is the target value, $y$ is the actual value of a product characteristic, $k$ is a constant, and $L$ is the financial loss.

A Taguchi quality loss curve is shown in Figure 4.18 (b). The curve describes how a loss, in terms of cost or financials, is greater as a product quality deviates more from the target value. In actual cases, the loss curve may be of different shape, and more complex than a parabolic curve, but the concept remains true. Therefore, the Taguchi quality loss function suggests that t, LSL, and USL should be based on customer satisfaction, and linked with financial cost or per-formance, considering variations.

The Taguchi loss function is a basic construct for all three pillars of a quality management system: quality planning, quality control, and continuous improvement, discussed in Chapter 1. Based on the same principle, the Taguchi loss function can also deal with two other situations within a one-sided specification: the lower the better for a USL only, and the larger the better for an LSL only.

For continuous improvement and variation reduction, the Taguchi loss func-tion implies a guideline: to find and prioritize attacking the variation that most affects customers. Some types of variation are minor or even innocuous to cus-tomers; combating these types of variation has no significant benefit, while add-ing tangible cost to customers. This guideline is in line with the customer-centric mindset from Chapter 3.

Recent research has focused on modeling loss curves for different applica-tions and introducing new methods. For example, one study was on hierarchi-cal products (Zhang et al. 2019). Another study jointly used the Taguchi loss function and FMEA (Sutrisno et al. 2018). An additional study integrated the Taguchi loss function and economic statistical design (Pasha et al. 2018).

# Summary

### Design Review Process

1 The objectives of a design review include conducting evaluations, testing functions to find solutions, and recommending adjustments to ensure design goals.

2 There are various types of design reviews, e.g. state-of-the-art review, requirement review, design kickoff review, system design review, internal/ critical design reviews, final verification review, etc.

3 DRBFM is a review approach that addresses the potential problems in a new or modified design, which is similar to design FMEA.

4 DRBFM has three steps: 1) good design, 2) good discussion, and 3) good dissection.

### Design Verification and Validation

5 A prototype is a working model built for a tangible or visible presentation; for demonstration, valuation, and/or validation purposes.

6 There are various types of prototypes, e.g. concept, rapid build, mockup, test prototypes, etc.

7 Verification, as an internal process, is to confirm that the specified requirements of a product or service have been fulfilled.

8 Validation is a process to confirm that the specifications of a product or service match external customer needs and intended uses at a system level.

9 Verification and validation processes are confirmative, and associated with design reviews and tests.

### CE

10 CE is a systematic approach that organically and simultaneously integrates related development tasks for new product and service development.

11 Cross-functional teams and a knowledge-sharing environment are necessary conditions for CE applications.

12 Core development teams, e.g. product design, play a central role in applying CE, and integrate other disciplinary teams for a particular product or service development.

13 The factors of successful CE applications include time commitment, corporate culture, computer environment, procedure, knowledge, etc.

**Variation Considerations**

14  Variation always exists in all types of processes and entities.
15  Sources of variation are of two general types: common causes and special causes.
16  Quality planning focuses on the common-cause variation of a product or service.
17  The target setting of a new product or service should include not only multiple influencing factors but also their variations.
18  The variations of individual components transfer to the complete system (product, service, or process). The transfer function is normally unknown, and is worth studying.
19  The Taguchi quality loss function reveals a relationship between quality changes and customer dissatisfaction.

# Exercises

## Review Questions

1  Describe the overall design review process.
2  Explain a type of design review with an example.
3  Discuss the inputs and outputs of a design review with an example.
4  Explain a characteristic of design review with an example.
5  Describe the process of DRBFM.
6  Use an example to show an acceptance criterion for a design review.
7  Discuss the differences between verification and validation of a product or service development.
8  Use an example to show a prototype process (e.g. concept, rapid build, mockup, or test prototypes).
9  Review an assessment of verification and validation with an example.
10  Discuss why verification and validation processes are confirmative, rather than exploratory or creative.
11  Describe the principle of CE.
12  Explain the benefits and challenges of CE applications.
13  Discuss the role of cross-functional teamwork in CE applications.
14  Use an example to review the necessity and feasibility of applying CE.
15  Describe the nature and meaning of variation with an example.
16  Explain the differences between common causes and special causes with examples.
17  A colleague of yours states that an operator's lack of training is a common cause of variation. Do you agree? Why?

18 Discuss why the variation from common causes, rather than special causes, should be addressed in quality planning.
19 Briefly explain how variation may be lessened or amplified when transmitting from a component to an assembly (or operation).
20 Compare the conventional ways and the Taguchi loss function in terms of judging quality.

## Mini-project Topics

1 Study the review exercises in a product or service development and comment on their necessity and functionality in quality assurance.
2 Locate a paper on a DRBFM application from library databases, or https://scholar.google.com, and discuss the characteristics of DRBFM.
3 "One states that DRBFM may discourage a new design or innovation." Please review this statement and provide justification for or against it.
4 Locate a paper on verification and validation from library databases, or https://scholar.google.com, and discuss the specific application of verification and validation.
5 Search a paper on the application of CE from library databases, or https://scholar.google.com, and discuss the differences from conventional engineering processes.
6 Find a product or service with an extremely high level of quality and review its variation.
7 One states that the focus of variation reduction is on common causes during planning and design while on special causes during execution and operation. Review and explain why you agree or disagree.
8 Find a paper on the quality target setting from library databases, or https://scholar.google.com, and discuss how to consider variation in the target setting.
9 Do a simple case study on variation transmissibility.
10 Find and review a case not using the Taguchi loss function, and discuss possible outcomes of using the Taguchi loss function.

# References

Abbasi, M. and Nilsson, F. (2016). Developing environmentally sustainable logistics: Exploring themes and challenges from a logistics service providers' perspective, Transportation Research Part D, *Transport and Environment*, Vol. 46, pp. 273–283. 10.1016/j.trd.2016.04.004.

Ajewole, T.O., Alawode, K.O., Omoigui, M.O. et al. (2017). Design validation of a laboratory-scale wind turbine emulator, *Cogent Engineering*, Vol. 4, Iss. 1, 10.1080/23311916.2017.1280888.

Albanese, R., Crisanti, F., Martin, P. et al. (2019). *Design Review for the Italian Divertor Tokamak Test Facility*, Vol. 146, Part A, pp. 194–197. 10.1016/j. fusengdes.2018.12.016.

Alkureidi, H. (2020). Investigations on Industrial Practice of Virtual Commissioning, Dissertation, University of Skövde, http://urn.kb.se/resolve?urn =urn:nbn:se:his:diva-18658, accessed in July 2020.

Andeson, D. (2020). *Design for Manufacturability: How to Use Concurrent Engineering to Rapidly Develop Low-Cost, High-Quality Products for Lean Production*, 2nd edition. New York, NY: Productivity Press.

Arnette, A. and Brewer, B. (2017). The influence of strategy and concurrent engineering on design for procurement, *International Journal of Logistics Management*, Vol. 28, Iss. 2, pp. 531–554. 10.1108/IJLM-03-2016-0081.

Başol, M.C., Pepeç, E., Kaya, Y. et al. (2019). Electronic design verification tests for LED driver: Analysis and implementation of worst case tests, *Procedia Computer Science*, Vol. 158, pp. 116–124. 10.1016/j.procs.2019.09.034.

CFR. (2012). 21 CFR 820.3–Definitions, CFR – Code of Federal Regulations Title 21, https://www.govinfo.gov/app/details/CFR-2012-title21-vol8/CFR-2012-title21-vol8-sec820-3, Accessed in April 2020.

Deming, W.E. (1994). *The New Economics: For Industry, Government, and Education*, 2nd edition. Cambridge, MA: the MIT Press.

Deshpande, A. (2019). Total quality management, concurrent engineering and manufacturing performance: An empirical investigation, *Journal of Operations and Strategic Planning*, Vol. 2, Iss. 1, pp. 35–64. 10.1177/2516600X19845230.

DOE. (n.d.). Commercial prototype building models, Office of Energy Efficiency & Renewable Energy, https://www.energycodes.gov/development/commercial/prototype_models, accessed in April 2020.

Eckert, P. (2018). *Meaning and Linguistic Variation : The Third Wave in Sociolinguistics*. Cambridge; Cambridge University Press.

FDA. (2018a). The device development process, US Food and Drug Administration, https://www.fda.gov/patients/learn-about-drug-and-device-approvals/device-development-process,Contentcurrentasof01/04/2018, accessed in April 2020.

FDA. (2018b). Process validation: General principles and practices - guidance for industry, US Food and Drug Administration, https://www.fda.gov/regulatory-information/search-fda-guidance-documents/process-validation-general-principles-and-practices,Contentcurrentasof08/24/2018, accessed in September 2020.

Filson, A. and Lewis, A. (2000). Cultural issues in implementing changes to new product development process in a small to medium sized enterprise (SME), *Journal of Engineering Design*, Vol. 11, Iss. 2, pp. 149–157.

Fischer, P.M., Deshmukh, M., Maiwald, V. et al. (2017). Conceptual data model: A foundation for successful concurrent engineering, *Concurrent Engineering*, Vol. 26, Iss. 1, pp. 55–76. 10.1177/1063293X17734592.

Galigekere, V.P., Pries, J., Onar, O. et al. (2018). Design and implementation of an optimized 100 kW stationary wireless charging system for EV battery recharging. *IEEE Energy Conversion Congress and Expo (ECCE 2018)*, Portland, Oregon. 10.1109/ECCE.2018.8557590. https://www.osti.gov/servlets/purl/1495980, accessed in August 2020.

Giganti, M.J., Shepherd, B.E., Caro-Vega, Y. et al. (2019). The Impact of Data Quality and Source Data Verification on Epidemiologic Inference: A Practical Application Using HIV Observational Data, *BMC Public Health*, Vol. 19, pp. 1–11. 1748. 10.1186/s12889-019-8105-2.

Hallgrimsson, B. and Hall, B. (Eds). (2011). *Variation: A Central Concept in Biology*. Amsterdam: Academic Press.

Hansen, C., Gosselin, F., Mansour, K.B. et al. (2018). Design-validation of a hand exoskeleton using musculoskeletal modeling, *Applied Ergonomics*. Vol. 68, pp. 83–288. 10.1016/j.apergo.2017.11.015.

Huang, R. and Li, Y. (2017). *Teaching and Learning Mathematics through Variation: Confucian Heritage Meets Western Theories*. Rotterdam: Sense Publishers.

IEEE, (2017). IEEE 1012–2016 IEEE sandard for system, software, and hardware verification and validation, *The Institute of Electrical and Electronics Engineers*, ISBN: 978–1504418126.

ISO. (2018). *ISO ISO 14064-1:2018 Greenhouse Gases – Part 1: Specification with Guidance at the Organization Level for Quantification and Reporting of Greenhouse Gas Emissions and Removals*. Geneva: International Organization for Standardization.

Kukulies, J. and Schmitt, R. (2018). Stabilizing production ramp-up by modeling uncertainty for product design verification using Dempster–Shafer theory, *CIRP Journal of Manufacturing Science and Technology*, Vol. 23, pp. 187–196. 10.1016/j.cirpj.2017.09.008.

Lechler, T., Fischer, E., Metzner, M. et al. (2019). Virtual Commissioning–scientific review and exploratory use cases in advanced production systems, *52nd CIRP Conference on Manufacturing Systems, Procedia CIRP*, Vol. 81, pp. 1125–1130. 10.1016/j.procir.2019.03.278

Liu, Y., Messner, J. and Leicht, R. (2018). A process model for usability and maintainability design reviews, *Architectural Engineering and Design Management*, Vol. 14, Iss. 6, pp. 457–469. 10.1080/17452007.2018.1512042.

Loannou, A., Papastavrou, E., Avraamides, M.N. et al. (2020). Virtual reality and symptoms management of anxiety, depression, fatigue, and pain: A systematic review, *SAGE Open Nursing*, Vol. 6, pp. 1–13. 10.1177/2377960820936163.

Loureiro, G., Panades, W.F. and Silva, A. (2018). Lessons learned in 20 years of application of systems concurrent engineering to space products, *Acta Astronautica*, Vol. 151, pp. 44–52. 10.1016/j.actaastro.2018.05.042.

Mackel, C. and Kinneir, L. (2019). Design review in Northern Ireland: The MAG process, *Journal of Urban Design*, Vol. 24, Iss. 4, pp. 613–616. 10.1080/13574809.2019.1614846.

Martin, P. (2019). The new divertor tokamak test facility project, presentation at plasma science and fusion center, Massachusetts Institute of Technology, January 23, 2019. http://www.psfc.mit.edu/events/2019/the-new-divertor-tokamak-test-facility-project, accessed in April 2020.

Mclaughlin, C.P. (1996). Why variation reduction is not everything: A new paradigm for service operations, *International Journal of Service Industry Management*, Vol. 7, Iss. 3, pp. 17–30. 10.1108/09564239610122938.

Mehrpouyan, H., Giannakopoulou, D., Brat, G. et al. (2016). Complex engineered systems design verification based on Assume-guarantee reasoning, *Systems Engineering: The Journal of the International Council on Systems Engineering*, Vol. 19, Iss. 6, pp. 461–476. 10.1002/sys.21368.

Mejía-Gutiérrez, R. and Carvajal-Arango, R. (2017). Design Verification through virtual prototyping techniques based on Systems Engineering, *Reliability Engineering Design*, Vol. 28, pp. 477–494. 10.1007/s00163-016-0247-y.

Mobin, M., Li, Z., Hossein Cheraghi, S. et al. (2019). An approach for design verification and validation planning and optimization for new product reliability improvement, *Reliability Engineering & System Safety*, Vol. 190, 106518. 10.1016/j.ress 2019.106518.

North, R., Pospisil, C., Clukey, R. et al. (2019). Impact of human factors testing on medical device design: Validation of an automated CGM sensor applicator, *Journal of Diabetes Science and Technology*, Vol. 13, Iss. 5, pp. 949–953. 10.1177/1932296819831071.

Pasha, M.A., Moghadam, M.B., Fani, S. et al. (2018). Effects of quality characteristic distributions on the integrated model of Taguchi's loss function and economic statistical design of $\bar{X}$-control charts by modifying the Banerjee and Rahim economic model, *Communications in Statistics–Theory and Methods*, Vol. 47, Iss. 8, pp. 1842–1855. 10.1080/03610926.2017.1328512.

Power, J.D. (2020). New-vehicle quality mainly dependent on Trouble-free technology, J.D. Power Finds, Press Release, June 24, 2020, https://www.jdpower.com/business/press-releases/2020-initial-quality-study-iqs, accessed in October 2020.

Putnik, G.D. and Putnik, Z. (2019). Defining Sequential Engineering (SeqE), Simultaneous Engineering (SE), Concurrent Engineering (CE) and Collaborative Engineering (ColE): On similarities and differences, *Procedia CIRP 84*, pp. 68–75, 29th CIRP Design, Póvoa de Varzim, Portugal, May 8–10, 2019. 10.1016/j.procir.2019.07.005

SAE, (2013). Design Review Based on Failure Modes (DRBFM), SAE J2886-201303, 10.4271/J2886_201303.

Sanjuan, J.D., Castillo, A.D., Padilla, M.A. et al. (2020). Cable driven exoskeleton for upper-limb rehabilitation: A design review, *Robotics and Autonomous Systems*, Vol. 126, 103445. 10.1016/j.robot.2020.103445.

Sehat, A.R. and Nirmal, U. (2017). State of the art baby strollers: Design review and the innovations of an ergonomic baby stroller, *Cogent Engineering*, Vol. 4, Iss. 1, 1333273. 10.1080/23311916.2017.1333273.

Sobek, D.K., Ward, A. and Liker, J. (1999). Toyota's principles of Set-based concurrent engineering, *MIT Sloan Management Review*, Vol. 40, Iss. 2, pp. 67–83.

Stjepandić, J., Verhagen, W., and Wognum, N. (2015). CE challenges – work to do, *Proceedings of the 22nd ISPE Inc International Conference on Concurrent Engineering*, July 20–23, 2015. 10.3233/978-1-61499-544-9-627.

Sutrisno, A., Gunawan, I., Vanany, I. et al. (2018). An improved modified FMEA model for prioritization of lean waste risk, *International Journal of Lean Six Sigma*, Vol. 11, Iss. 2, pp. 233–253. 10.1108/IJLSS-11-2017-0125.

Union, A.H., Kadhm, H., Mahar, A. et al. (2020). The prospect of using concurrent engineering for enhancing supply chain efficiency and reducing costs in the hospitality sector, *African Journal of Hospitality, Tourism and Leisure*, Vol. 9, Iss. 2.

Weber, S. and Duarte, C. (2017). *Circuit Design – Anticipate, Analyze, Exploit Variations: Statistical Methods and Optimization*. Gistrup: River Publishers.

Wright, D. (2011). Application of Mizenboushi ($GD^3$) method of problem prevention to vehicle, component and subsystem validation. SAE Technical Paper 2011-04-12, *SAE 2011 World Congress & Exhibition*. 10.4271/2011-01-1275.

Yamaguchi, T., Brain, M., Ryder, C. et al. (2019). Application of abstract interpretation to the automotive electronic control system, in Enea, C. and Piskac, R. (Eds), *Verification, Model Checking, and Abstract Interpretation. VMCAI 2019. Lecture Notes in Computer Science*, 11388, pp. 425–445., 10.1007/978-3-030-11245-5_20.

Zhang, Y., Li, L., Song, M. et al. (2019). Optimal tolerance design of hierarchical products based on quality loss function, *Journal of Intelligent Manufacturing*, Vol. 30, Iss. 1, pp. 185–192. 10.1007/s10845-016-1238–6.

Zhou, J. and Li, D., (2009). Reliability verification: plan, execution, and analysis, SAE Technical Paper 2009-01-0561. 10.4271/2009-01-0561.

# 5

# Proactive Approaches

**Failure Modes and Effects Analysis and Control Plan**

## 5.1 Understanding Failure Modes and Effects Analysis

### 5.1.1 Principle of Failure Modes and Effects Analysis

**Process of Failure Modes and Effects Analysis**

Any product or service can have various quality problems, such as defects, malfunctions, and delays, when customers use a product or take a service. Minimizing and preventing quality problems is a major task of quality planning in development.

Failure modes and effects analysis (FMEA) is a proactive approach to identify potential failures, assess their possible impacts, and recommend countermeasures in advance. The aims of doing FMEA are:

- To identify potential failure modes, and evaluate their possible impacts.
- To propose preventive actions, and prioritize these to mitigate higher risks of potential failures and reduce their impacts.
- To follow up on implementation plans to ensure improved quality of a product or service.

The FMEA approach can apply to a system, such as a product, service, process, or other business function. FMEAs are normally conducted for new and high-risk subsystems and main functions of a large or complex system. A specific FMEA may be dedicated to a subsystem, such as a design FMEA, service FMEA, or process FMEA. For a new product, both a design FMEA and process FMEA are normally required.

Design FMEA is very similar to the DRBFM (design review based on failure mode) process discussed in Chapter 4, in terms of principle, process, and documentation. Most times, design FMEA and DRBFM are considered equivalent.

*Quality Planning and Assurance: Principles, Approaches, and Methods for Product and Service Development*, First Edition. Herman Tang.

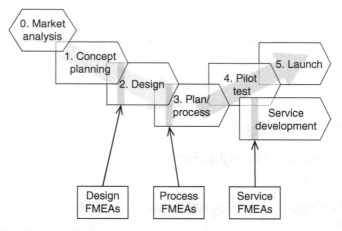

**Figure 5.1** Timing of FMEA development.

Being predictive in nature, FMEA development should be in advance of specific development. For example, for a product to be developed with a corresponding service, the design, process, and service FMEAs should be planned and conducted in the early stages of respective phases (Figure 5.1). The FMEAs should also be followed up in later phases.

As a process, FMEA requires various inputs and provides the outputs of analysis results and recommendations, following predefined processes and criteria, as shown in Figure 5.2. The inputs and outputs are discussed in the following subsections.

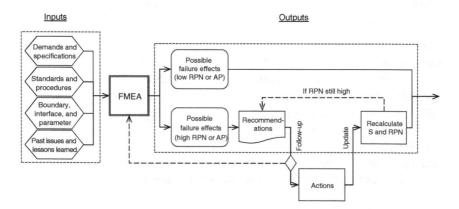

**Figure 5.2** Inputs and outputs of FMEA.

**Input of FMEA**

FMEA development, as a structured analysis approach, relies on a lot of existing information, knowledge, and experience. The <u>inputs</u> and associated considerations of FMEA include:

- Specifications of products, services, and/or processes, relative to customer quality expectations and to the current capacity to meet them. For example, if a new specification is higher than the existing ones, it needs special attention in FMEA.
- Relevant standards and procedures must be in place. They will be used to evaluate potential failures and recommend countermeasures. Insufficient standards and procedures can slow FMEA development and/or render FMEA incomplete.
- FMEA procedure and predefined criteria for a given case must be ready. The training of participants and readiness of the FMEA criteria are prerequisites to its success.
- Analysis tasks may be needed to prepare and support FMEA development. Such analyses may include a boundary diagram, interface matrix, and parameter diagram, which are discussed in section 5.2. The results from other studies may be used as well, e.g. risk assessment (Chapter 2), quality function deployment (QFD) (Chapter 3), design review (Chapter 4), and/or process capability study (Chapter 7).
- Past quality issues and lessons learned are important references. Likewise, the issues found from benchmark and competition analyses on similar products or services are useful references.

FMEA should be conducted by a cross-functional team in the development phase to determine appropriate actions that will make a product or service better. FMEA is also a brainstorming exercise. The knowledge and experience of the cross-functional team members are crucial to the value and success of FMEA. Depending on FMEA targets, the FMEA team members may include quality, operations, engineering, manufacturing, maintenance, sales, and/or marketing, etc. They should be knowledgeable on the basic aspects of a future product or service, such as the incoming materials and existing operations.

**Output of FMEA**

The <u>outputs</u> of FMEA teamwork are a completed FMEA document, or FMEA form. A typical form is shown in Table 5.1, and actual examples are given in Tables 5.9–5.12, to be discussed in later subsections. The information of an FMEA document includes:

- All potential failure modes of a product, service, or process.
- All possible impacts from the failure modes identified.
- Assigned rating values of the severity, likelihood, and detectability of each failure mode.
- A risk priority number (RPN) value for each failure mode, or an action priority (AP) list.

**Table 5.1** Common FMEA format with administrative (head) section.

| Title: | | Area: | | Owner: | | Initial date: |
|---|---|---|---|---|---|---|
| Author: | | Team: | | | | Rev. date: |

| 1. Feature/ function, team | 2. Potential failure mode | 3. Potential effect of failure | 4. Failure severity rating ($S_1$) | 5. Possible cause of failure | 6. Occurrence rating ($O_1$) | 7. Current control method | 8. Failure detection rating ($D_1$) | 9. $RPN_1$ ($=S_1 \times O_1 \times D_1$) | 10. Recommended action | 11. Responsibility and date | 12. Action taken | 13. Post-action severity ($S_2$) | 14. Post-action occurrence ($O_2$) | 15. Post-action detection ($D_2$) | 16. Post-action $RPN_2$ |
|---|---|---|---|---|---|---|---|---|---|---|---|---|---|---|---|
| Step 1 | Step 2 | | | | | Step 3 | | | Step 4 | | | | Step 5 | | |

- The re-evaluated RPN or AP list of these respective failure modes, based on the recommended actions taken. Note this item is available after the recommended actions have been taken.

An AP list of recommended actions is an alternative to an RPN. In an AP list, three levels of priorities are identified: high (must do), medium (should do), and low (could do). An AP decision criteria reference is shown in Figure 5.3.

The core values of a completed FMEA are its predictive findings and prioritized recommendations. The recommendations based on RPNs or AP list can be guidelines to modify or redesign a product, service, or process. To claim full success in FMEA work, one should follow the recommendations, take deliberate actions, and update FMEA documents to ensure all major identified issues are resolved. The actions should result in reduced severity, reduced likelihoods, and/or increased detectability of the failure modes. The recalculated RPNs should be smaller, or the

| ① Severity | ② Occurrence | ③ Detection | Priority |
|---|---|---|---|
| 9-10 (Catastrophic) | 8-10 (Certain) | 1-10 (regardless) | High |
| | 6-7 (Possible) | 7-10 (Slight) | High |
| | | 5-6 (Moderate) | High |
| | | 2-4 (Adequate) | High |
| | | 1 (Excellent) | High |
| | 4-5 (Moderate) | 7-10 (Slight) | High |
| | | 5-6 (Moderate) | High |
| | | 2-4 (Adequate) | High |
| | | 1 (Excellent) | Medium |
| | 2-3 (Remote) | 7-10 (Slight) | High |
| | | 5-6 (Moderate) | Medium |
| | | 2-4 (Adequate) | Low |
| | | 1 (Excellent) | Low |
| | 1 | 1-10 (regardless) | Low |
| 7-8 (Moderate) | 8-10 (Certain) | 7-10 (Slight) | High |
| | | 5-6 (Moderate) | High |
| | | 2-4 (Adequate) | High |
| | | 1 (Excellent) | High |
| | 6-7 (Possible) | 7-10 (Slight) | High |
| | | 5-6 (Moderate) | High |
| | | 2-4 (Adequate) | High |
| | | 1 (Excellent) | Medium |
| | 4-5 (Moderate) | 7-10 (Slight) | High |
| | | 5-6 (Moderate) | Medium |
| | | 2-4 (Adequate) | Medium |
| | | 1 (Excellent) | Low |
| | 2-3 (Remote) | 7-10 (Slight) | High |
| | | 5-6 (Moderate) | Medium |
| | | 2-4 (Adequate) | Low |
| | | 1 (Excellent) | Low |
| | 1 | 1-10 (regardless) | Low |

**Figure 5.3** Sample of AP criteria of FMEA.

high priority issues on the AP list should be resolved. In some cases, a final executive summary is created after FMEAs are completed and closed.

### 5.1.2 FMEA Development

#### Overall Development Process

There are several published standards and guidelines. They include ISO 12132:2017 for plain design (ISO 2017), IEC 60812:2018 for system reliability (IEC 2018), SAE J1739 (SAE 2009) and FMEAAV-1 (AIAG and VDA 2019) for the automotive industry, ARP5580 for nonautomotive applications (SAE 2012), and ARP4761 for the civil aviation industry (SAE 1996). In service sectors, several institutes provide their own FMEA guidelines. These standards and guidelines are similar in their formats, items, titles, etc.

To develop an FMEA, one takes five steps (Figure 5.4) and completes the corresponding items in an FMEA template (Table 5.1). Note that the latest AIAG–VDA FMEA guideline suggests seven steps, which adds complexity, requires more time, and might reduce FMEA accuracy and completeness (Kluse 2020).

It is a good practice for management to review and sign off after Step 4, to ensure an FMEA is complete and satisfactory, shown as "◊" in Figures 5.2 and 5.4. A management review, as a type of audit, can be effective to prevent incomplete or items missing in an FMEA and improve its quality.

*Step 1. Preparation*

The first step is to define the scope of the FMEA for a product, process, or service, identify its functions, and collect relevant information (item 1 in Table 5.1). For a large FMEA, one may need to break it into subsystems and key elements. In this step, an FMEA team is formed with a responsible person (an "owner") and cross-functional team members. Three important tasks for the team are:

- To "walk" with the FMEA target, e.g. work with a process flow.
- To understand and define the scope of work.
- To collect all necessary information and inputs pursuant to the scope of work.

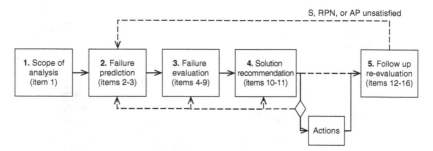

**Figure 5.4** Overall process flow of FMEA development.

The first step may take some time, for information collection, to ensure the readiness of FMEA development. Some pre-FMEA work needs to be done, which is discussed in section 5.2.1.

*Step 2. Failure Prediction*

In this step, the FMEA team examines the major functions or process steps of a new product or service, and brainstorms for potential issues to identify:

- Potential failure modes (item 2)
- Potential effects of the failure modes (item 3)

Note that a common practice is to consider normal operations, rather than extreme situations in FMEA development. Extreme situations or major external factors should be considered in design if they have significant influence, such as a power loss might influence a surgical operating room. Most times, however, unintended usages are included in FMEA. Furthermore, the interactions between an FMEA target and its interfaces should be considered as well, which is discussed in section 5.2.1.

*Step 3. Failure Evaluation*

In the third step, the team evaluates the potential failure modes by referring to historical data and predefined ratings and analyzes the following items (subscript "1" indicates the initial state):

- Severity ratings ($S_1$) of the effects of each failure mode (item 4)
- Possible root causes of each failure mode (item 5)
- Occurring likelihood rating ($O_1$) of each failure mode (item 6)
- Current control method to prevent each failure mode (item 7)
- Detection possibility rating ($D_1$) of each failure mode (item 8)
- Calculation of $RPN_1 = S_1 \times O_1 \times D_1$ (item 9)

As a ranking number, an RPN itself has no physical meaning, but serves as an indicator to the priorities of risks of failure.

*Step 4. Making Recommendations*

Action recommendations are based on either RPN or AP. An RPN treats S, O, and D equally, which may be debatable because they can have different levels of significance to an application. In an AP process, severity is considered first, then the occurrence and so on. As both RPN and AP are based on the three parameters of S, O, and D, resulting recommendations can be similar in purposes:

- Improvement or countermeasure actions, in terms of a design change, replanning, or additional feature, etc. to reduce the RPN and/or S rating of the failure modes concerned (item 10).
- Responsible departments to implement the recommended actions before the due dates (item 11).

If there are no recommended actions for certain potential failure modes, one should use "None" to acknowledge that the step was not skipped but actions are not needed at this time. Improvement is an iterative process and after, say, the top three actions are implemented, one may go back and consider the next three potential failure modes with the highest RPNs.

The recommendations are the value outputs of an FMEA. An FMEA that has many recommended actions may be a strong FMEA, meaning the team has found various ways to improve the product or service. If an FMEA does not come with any recommendations, it implies that the FMEA may be unnecessary or not conducted thoroughly.

*Step 5. Re-evaluation after Actions*

The FMEA team re-evaluates and gets the $S_2$, $O_2$, and $D_2$ ratings (subscript "2" shows their updated states) and recalculates RPNs of the failure modes, after the recommended actions are implemented:

- Documentation of the completed improvement actions (item 12).
- Re-evaluation of $S_2$, $O_2$, and $D_2$ ratings, and determine if $S_2$ is satisfactory (items 13–15).
- Recalculation of $RPN_2$, and determine if it is satisfactory or lowered on the AP list (item 16).

With the completion of FMEA, the potential failure modes with a high RPN, or on an AP list, or just high severity rating should be significantly improved (or reduced). Note that a severity rating of a failure mode is determined by design, rather than an improvement in execution. In other words, the severity ($S_2$) rating, after a recommended action, can be changed *only* if the failure mode is changed by design modifications. Most improved RPNs have better $O_2$ and/or $D_2$ but with unchanged $S_2$.

### 5.1.3 Parameters in FMEA

In the development of FMEA, it is important to evaluate and rate potential failure modes on their S, O, and D values. These rating parameters are often on a scale of 1–10, where "10" is the worst and "1" is the best. It is vital that these rating parameters are well defined prior to working on the FMEA. Because rating values are directly related to the target (product, process, service), the discussion that follows is based on certain applications.

#### Severity Rating

In general, severity assesses how serious the effects would be should a potential risk occur. Severity level is rated against the impact of the effect caused by a failure mode. For example, the severity for a manufacturing process failure mode is rated based on potential safety issues and downtime. Table 5.2 shows as an example of severity rating for mass production manufacturing systems (Tang 2018).

**Table 5.2** Severity rating example for process FMEA

| Severity | Criteria (based on safety and downtime) | Rating |
|---|---|---|
| Hazardous w/o warning | Affecting personnel safety, not-compliance with government | 10 |
| Hazardous w/warning | Affecting personnel safety, not-compliance with government | 9 |
| Very high | Downtime > 8 h or defective parts > 4 h | 8 |
| High | Downtime > 4 h or defective parts > 2 h | 7 |
| Moderate | Downtime > 1 h or defective parts > 1 h | 6 |
| Low | Downtime > 30 min or defective parts found | 5 |
| Very low | Downtime > 10 min but no defective parts | 4 |
| Minor | Downtime < 10 min but no defective parts | 3 |
| Very minor | Quickly process adjustment needed during production | 2 |
| None | No action needed during production | 1 |

*Source*: Tang, H., (2018). *Manufacturing System and Process Development for Vehicle Assembly*, ISBN-13: 978-0-7680-8346-0, Warrendale, PA: SAE International.

The rating values for downtime duration are process specific. The downtime criteria should be shorter for a low-volume production. For example, a high severity rating of 7 may be defined as downtime > 1 h, instead of > 4 h, in the table. For some industries, there are regulations for the potential failure

**Table 5.3** Severity rating example for product FMEA

| Severity | Criteria (based on safety and downtime) | Rating |
|---|---|---|
| Hazardous without warning | Safety issue or noncompliance with government regulation | 10 |
| Hazardous with warning | Safety issue or noncompliance with government regulation | 9 |
| Very high | Loss of a major function | 8 |
| High | Reduction of a major function | 7 |
| Moderate | Loss of a minor function, e.g. comfort or convenience | 6 |
| Low | Reduction of a minor function, e.g. comfort or convenience | 5 |
| Very low | Noticeable appearance or minor issue | 4 |
| Minor | Minor appearance | 3 |
| Very minor | Non-noticeable appearance | 2 |
| None | No issue | 1 |

evaluations. For example, the impact of equipment or systems on the safe operation of an airplane is rated catastrophic, hazardous, major, or minor with detailed definitions (FAA 2011). Financial impact, e.g. loss, may be used as a criterion for a severity rating as well. Table 5.3 shows a general reference of a severity rating for a product design FMEA. For a specific product, this general rating reference needs to be clearly defined.

### Occurrence and Detection Ratings

Occurrence likelihood evaluates the frequency that potential failure mode will occur for a given situation. One may decide ratings based on occurrence frequency. For manufacturing systems, one can use the mean time between failures (MTBF) as criteria. Table 5.4 lists two examples for manufacturing systems of a high volume production (Tang 2018). Occurrence rating can be based on historical data, if it is available.

Detection rating is defined similarly. In general, detectability is the probability of a failure being immediately detected when it occurs in a system. One may define ratings for failure detectability, where the higher chance to detect is better. Historical data is also a reliable source to define detection rating. Table 5.5 lists an example of detectability for a failure mode, if occurring.

**Table 5.4**  Occurrence rating example for process FMEA

| Probability | Criteria 1: MTBF (h) | Criteria 2: Frequency | Rating |
| --- | --- | --- | --- |
| Very High | MTBF < 1 | >10% | 10 |
| Very High | 2 < MTBF < 10 | >5% | 9 |
| High | 11 < MTBF < 100 | >2% | 8 |
| High | 101 < MTBF < 400 | >1% | 7 |
| Moderate | 401 < MTBF < 1,000 | >0.5% | 6 |
| Moderate | 1001 < MTBF < 2,000 | >0.2% | 5 |
| Moderate | 2001 < MTBF < 3,000 | >0.1% | 4 |
| Low | 3001 < MTBF < 6,000 | >0.05% | 3 |
| Low | 6001 < MTBF < 10,000 | >0.01% | 2 |
| Remote | MTBF > 10,000 | >0.001% | 1 |

*Source*: Tang, H., (2018). *Manufacturing System and Process Development for Vehicle Assembly*, ISBN-13: 978-0-7680-8346-0, Warrendale, PA: SAE International.

**Table 5.5** Detection rating example for process FMEA

| Detectability | Possibility (equivalent %) | Rating |
|---|---|---|
| Almost certain | 1 in 1 (~100%) | 1 |
| High | 1 in 2–10 (10–50%) | 2–3 |
| Medium | 1 in 10–100 (1–10%) | 4–5 |
| Slight | 1 in 100–500 (0.5–1%) | 6–7 |
| Probably not | 1 in 500 1,000 (0.1–0.5%) | 8–9 |
| Almost impossible | 1 in > 1,000 (< 0.1%) | 10 |

### For Different Applications

For different types of FMEA applications, such as a design, product, or service function, these rating parameters can be very different, in terms of physical indicators and their values. Some industries and organizations have their own standards to define ratings. For example, NASA (2010) defines the severity calculation for ground support equipment.

In many cases, ratings are particular to specific products, e.g. video games (Tanka et al. 2014) and airbags (Aljazzar et al. 2009). Table 5.6 shows two examples of FMEA severity ratings: one for an aging care service (Chen 2016) and the other as a general case (McCain 2006), respectively. A service FMEA was based on a questionnaire survey for a hypermarket, where the severity was on a scale

**Table 5.6** Examples of severity ratings for service FMEA

| Criteria 1 | Ranking | Criteria 2 | Ranking |
|---|---|---|---|
| Life and safety of customer are affected | 9–10 | Noncompliance with government regulations | 9–10 |
| Customer is tremendously unsatisfied | 7–8 | Downtime, large costs incurred | 7–8 |
| Customer is not satisfied | 4–6 | Significant customer rejection | 5–6 |
| Customer is slightly troubled | 2–3 | Confrontation, additional cost | 3–4 |
| Customer may not pay attention | 1 | Dissatisfied customer still using service | 2 |
|  |  | No effect | 1 |

of 1–5, where "5" meant the most severe (Chuang 2007). Interestingly, the severity ratings in such an FMEA are decimal numbers, such as 2.83 as an average of survey respondents' ratings.

## 5.2 Pre- and Post-work of FMEA

### 5.2.1 Pre-FMEA Analysis

Some analyses are necessary prior to FMEA, as their results serve as inputs and references for the FMEA. Completing one or all of these analyses can help improve FMEA work efficiency by reducing unnecessary discussion on the scope of FMEA development, among other things.

**Boundary Diagram**

When starting FMEA, it is possible that there is not a clear definition and scope of work. In such cases, one may use a boundary diagram to define the target system (product, service, process, etc.) to be analyzed. As a graphical representation, a boundary diagram helps an FMEA team understand the internal subsystems, their relations, their interfaces to external systems, etc. Figure 5.5

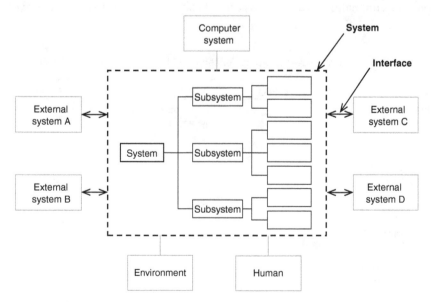

**Figure 5.5** Boundary diagram for FMEA.

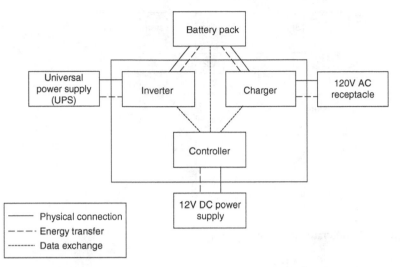

**Figure 5.6** Example of an FMEA boundary diagram. *Source:* Catton, W.A., Walker, S.B., McInnis, P. et al. (2019). Design and analysis of the use of re-purposed electric vehicle batteries for stationary energy storage in Canada, Batteries, Vol. 5, No. 14, 19 pages, 10.3390/batteries5010014. Licensed under CCBY 4.0.

illustrates a general structure of a boundary diagram. Figure 5.6 is an example of an FMEA boundary diagram for a vehicle battery system (Catton et al. 2019).

For FMEA, a boundary diagram draws the borderline of a system to be addressed. However, a boundary diagram does not suggest that the FMEA only includes the internal elements of a target system. An FMEA team should also consider the interfaces between this internal system and the surrounding external systems, identified by a boundary diagram, because failures may occur on these interfaces. For example, floor mats interfered with accelerator paddles in vehicles, which led to traffic accidents (NBC 2006; James 2009).

### Interface Matrix

A complex system has multiple subsystems, with certain relationships among them. It is important to understand these relationships before doing an FMEA. An interface matrix may be used to study and show these relationships and interfaces as a supplement to a boundary diagram, as the matrix addresses these specific relationships.

Figure 5.7 shows the typical structure of an interface matrix. In the matrix, the relationship between two systems may have multiple characteristics: C1, C2, etc. The typical relationships include data, material, physical, and spatial characteristics.

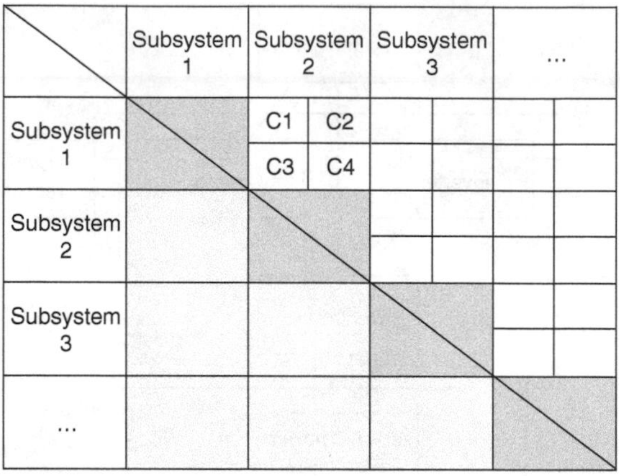

**Figure 5.7** Interface matrix for an FMEA.

Table 5.7 shows an example of an interface matrix for the power conditioner in a proton-exchange membrane fuel cell (Rastayesh et al. 2020). In the table, the interface characteristics are physical (P), material exchange (M), energy transfer (E), and data exchange (D).

In addition to a yes/no indication, the characteristics of an interface matrix may also be presented as a simple importance level, on a scale of 1–3. In most cases, interface matrixes are symmetrical along a diagonal line. In other words, the lower left part of an interface matrix is redundant. There wil be more discussion of an interface matrix or matrix chart in Chapter 8.

**Table 5.7** Example of an FMEA interface matrix

|  | Auxiliary power supply | Power stage | Controller | Gate driver | Printed circuit board |
|---|---|---|---|---|---|
| Auxiliary power supply |  | PED | PD | PED | P |
| Power stage | PED |  |  | PED | P |
| Controller | PD |  |  | PD | P |
| Gate driver | PED | PED | PD |  | P |
| Printed circuit board | P | P | P | P |  |

**Parameter Diagram**

A parameter diagram (or P-diagram) is a general system overview. A typical P-diagram, seen in Figure 5.8, visually shows the inputs to a system, desired outputs, and noncontrollable outside influences for the FMEA team. A P-diagram is useful for a complex system with many interactions, operating conditions, and parameters, e.g. high-value, low-volume manufacturing scenarios (Winter et al. 2015) and product design (Enoch et al. 2015).

The inputs of a system include the resources, materials, and data that make it work. The outputs include intended results through a given transformation of energy, material, or information, sometimes including unintentional outputs. Both expected and unintended outputs should be addressed in FMEA work.

Control factors are typically the design parameters that intentionally affect system performance, and are under the control of management or an engineering team, e.g. electrical requirements for a given appliance. A complex system can have hundreds of control parameters. In contrast, noise factors can be known or unknown unfavorable factors to the system outputs. It is vital to identify all noise factors and control their influences. Noise factors may include:

- Operation variation, i.e. piece-to-piece variation
- Normal degradation of materials or equipment over time, e.g. wear out and fatigue
- Environmental conditions, e.g. temperature and humidity
- System interactions with other systems

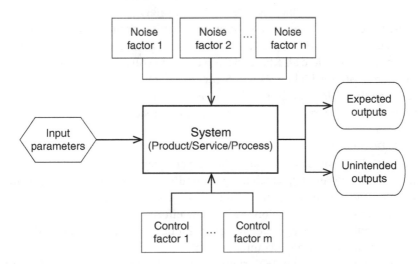

**Figure 5.8** P-diagram for FMEA.

## 5.2.2 FMEA Follow-up

### Follow-up Criteria

The value of FMEA is its function for predictions and recommendations on major potential issues. Regarding the outputs of an FMEA, follow-up actions are needed based on the recommendations on the high-severity (S) ratings, high-resultant RPNs, or high-priority items on an AP list.

A common practice is to look at the failure modes for S > 7 first. Concerns for high severity ratings are that their failure modes can be a safety hazard to customers, or do not comply with government regulations. Such failure modes should be prevented if possible, or otherwise dealt with. Ensure FMEA outputs with a potential failure of S > 7 are fed back to the design team. The function associated with a high-severity failure mode should be labeled as a critical characteristic. Such a characteristic should remain of critical status, even after the severity rating is reduced by a redesign or other action taken.

An RPN is another evaluation point for recommendations. Companies normally consider an RPN > 100 or 120 for direct actions, when all three ratings of S, O, and D are on a scale of 1–10. For some products or operations, where management wants to be risk-averse, the criteria may be lower, e.g. an RPN > 80. Review of occurrence and detection ratings for failure modes remain prudent regardless of unique benchmarks. For example, three failure modes have the same S rating (5) and RPNs (120). However, their O = 4, 6, and 8; D = 6, 4, and 3, respectively. A team review would decide which one should be addressed with a higher priority.

It is also important to realize that some failure modes cannot be 100% prevented, nor changed in their nature, such as a safety issue. In such cases, it is crucial to reduce occurrence, and improve detection significantly. One should target a much lower RPN, e.g. ≤ 40 with both O and D ≤ 2, for such failure modes.

In an instance where all RPN and S are low, one still ought to look at the top three items with the highest RPN or S, for continuous improvement opportunities. Sometimes, it is meaningful to review the borderline failure modes to ensure their S, O, and D ratings are appropriately assigned.

### Criteria Discussion

RPN and AP are subjective, because they are determined by the rating values of parameters (S, O, and D), which are profoundly experience-based. Influencing factors on the parameter ratings also include active participation, team brainstorming, available information, and review efforts. Therefore, resultant RPN and AP should be used as a reference point for decision-making, rather than as an absolute finality, in any planning scenario.

A good practice is that an RPN criterion is treated as a range, e.g. 100–150. This flexible criterion makes business sense, because follow-up actions require certain efforts and resources. One can use a Pareto chart to sort the RPNs for recom-

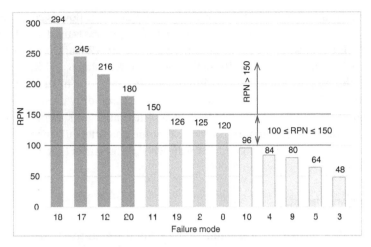

**Figure 5.9** Using a Pareto chart to select failure modes.

mendation consideration (refer to Figure 5.9 as an example; Pareto charts will be discussed further in Chapter 8). Here, the failure modes with an RPN > 150 must be addressed and ones with an RPN < 100 are fine. For the failure modes with an RPN of 100–150 (Figure 5.10), one needs to consider other factors, such as resources and project time, to decide how to deal with those failure modes.

Some organizations also consider individual occurrence and detection ratings, i.e. O or D > 7; meaning the likelihood of a failure occurrence is high, or failure detection is unlikely. For such failure modes, even though they have a low S and RPN, one needs to consider reducing their occurrence or improving detectability by design.

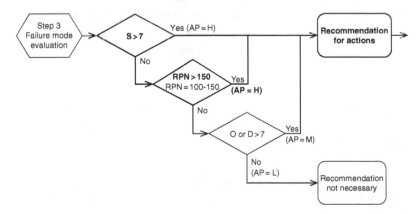

**Figure 5.10** Decision flow and criteria to recommend.

In consideration of all these criteria, an overall decision flow is summarized for recommendations in Figure 5.10, where the same process is used to determine the high (H) and/or medium (M) priority items on an AP list. Compared with using the AP process, many FMEA practitioners prefer using S and RPN.

When the detectability of most failure modes is similar, or in such cases less important, a simple way to rank failure modes is based only on their severity and occurrence, RPN' = S × O, which is sometimes called a risk score. This risk score RPN' may be easier to interpret on why some potential failure modes need attention. Clearly, the criterion based on an RPN' differs from that based on the conventional RPN.

### Follow-up and Update

An FMEA is an ongoing task over time, rather than a one-time event. An FMEA team and their management should follow up and revisit the FMEAs, update statuses, and decide to close if all the recommendations have been addressed. The "owner" of the product or service is normally responsible for managing FMEA follow-ups.

Recommended actions of FMEAs can lead to either a simple modification, or a new project. Sometimes, a follow-up project is conducted and documented as a control plan. The recommendations fall into the following categories.

1. To work on the root causes for removal. This requires modifications to design parameters, and maybe partial redesigns. These changes include optimization and using different technology.
2. To make the current design insensitive to the root causes. In this way, the root causes will less likely cause the failure modes of a product or service, or reduce the occurrences of these failure modes.
3. To compensate for a specific failure mode. The compensation may reduce the severity and/or occurrence of the failure modes.
4. To disguise the effects of failure modes. Such efforts include to improve detection of the failure mode or its effect, and to have the error or warning messages in place to reduce potential damage in advance, if no other option of improvement is available.

An FMEA team creates valuable recommendations. Their value can be realized only if these recommendations are implemented. For product or service development, implementation and update can have tight timing, e.g. a week. An FMEA should be updated when:

1. Main parameters are modified.
2. A new design or significant revision is introduced.
3. The original assumption or working environment is changed.
4. An undetected defective product reaches customers.

**Management Approval**

To ensure an appropriate follow-up and update, one should have FMEA follow-up processes and requirements in place. Figure 5.11 shows an FMEA follow-up process flow. Management review is part of FMEA development ("◊" in Figures 5.2 and 5.4). An FMEA can be considered completed only if all Ss, RPNs, or AP items are satisfactory, with final approval from management.

Most times, FMEA serves as a living technical document. One should revisit an "old" FMEA, and reassess the failure modes for any changed conditions, and for verifying the accuracy of previous rankings. For example, one revisits and updates an FMEA document with any new failure modes incurred, after the launch of a product or service. It is very possible that the actual failure modes differ from the predicted ones in an FMEA. For example, a failure mode had an RPN = 60 in the original FMEA (Cassanelli et al. 2006). After two years of production, the occurrence of the failure found increased, which resulted in a new RPN = 150. After a corrective action on the detection, the RPN was reduced to 100. With an additional action to reduce the occurrence, the RPN was further reduced to 40.

If a product or service is deemed mature, its FMEAs may be closed with two important steps. One is a follow-up review to ensure *all* failure modes with high RPNs and Ss are satisfactorily addressed. The representatives associated with the failure modes should participate in such a review. Then, the FMEA is submitted for management review and approval, to decide whether to close the FMEA or keep it open for a while. Many practices miss these two last steps, which results in outdated FMEA files "sleeping" in computers. Because of multiple failure modes being addressed in an FMEA, a verification review and management approval for closure may need to be conducted multiple times.

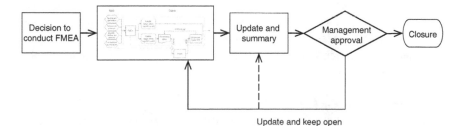

**Figure 5.11**   FMEA development and follow-up process flow.

## 5.3 Implementation of FMEA

### 5.3.1 Considerations in FMEA

**Team Commitment**

The principles and procedures of FMEA are not difficult, but their successful implementation requires dedicated teamwork. To be effective in brainstorming activities, an FMEA team should be cross-functional, and limited to five or six people. Wolniak (2019) showed that team member selection was a key factor to an FMEA success. If a team consists of people with the same background and function, then the outcomes of brainstorming activities can be limited.

The outputs from an FMEA should be based on team consensus, rather than majority rule or voting. This can be challenging because team members are from different disciplines. To reach a consensus, a team may require thorough brainstorming and discussion. For example, team members might need to take time to explain, debate, and/or justify the ratings they individually proposed. In such a discussion, it is important to avoid one person dominating or even dictating a team discussion, because of reporting relationship or personality. Concordantly, the facilitator of an FMEA should be able to navigate team dynamics.

A team consensus could be more difficult in service sectors than in engineering because the information in services is often subjective. In the healthcare industry, for example, the participants of a service FMEA study can have unique experiences and viewpoints on a current service, expectations for the future services, and severity of potential issues.

Time commitment from all members is crucial to the success of an FMEA because the work itself and reaching a team consensus can be time consuming. For example, the cross-functional teams of an automaker worked on process FMEAs on a weekly basis for six months. Thus, proper planning should be conducted allowing sufficient time to do FMEAs. Many companies have requirements on FMEA, but do not have the dedicated budget or time for FMEA development. Rushing on filling out an FMEA form, e.g. doing cut and paste from similar FMEAs, one person doing everything and then the team approving it, and preparing paperwork just to satisfy a customer or external audit, can cause an FMEA to be less valuable. As a result, an FMEA would serve merely as documentation, rather than a tool that can be used to predict and prevent problems and reduce risks.

**Data and Other Issues in FMEA**

FMEA is also data driven; thus obtaining sufficient amounts of information is essential. This information, from benchmarking, customer warranty reports, field complaints, other failure analyses, and previous lessons learned, etc., is important to predict and analyze potential failure modes. Because the design of

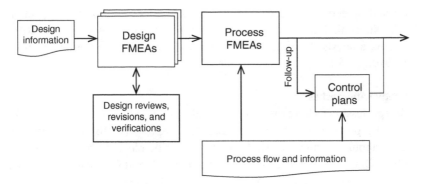

**Figure 5.12**   Relationship between design FMEA and process FMEA.

a product or service is prior to process planning in development, a design FMEA should be conducted first, and its outputs become the inputs to a process FMEA. For example, the major failure modes identified from a design FMEA, if not preventable, should be addressed in the corresponding process FMEA for detection and control, as shown in Figure 5.12. Manufacturing engineers determine and design the monitoring and/or control processes into manufacturing production for such design failure modes.

Not every FMEA is successful, even with well-documented files. One can justify FMEAs only based on their contributions to the success of a product or service. A few studies have been conducted on existing FMEA applications (Johnson and Khan 2003; Spreafico et al. 2017). The issues they found were related to the subjectivity, time consumption, understanding of cause-effect, and data management, etc. These studies indicated the difficulty in quantifying the true benefits of FMEAs in terms of costs, reliability improvements, and problem prevention.

Other pitfalls observed in FMEA development are listed here, from some industrial practices. Avoiding these issues that rely on system design, procedure, policy, management support, corporate culture, etc. requires in-depth thinking and continuous improvement.

- One-time documentation, without serious follow-ups and updates.
- Afterthought or after-the-fact "faux-proactive" efforts.
- Not being thorough, not addressing all high-risk potential failure modes.
- Missing critical or special/significant characteristics.
- Only focusing on the final product, not considering its main subassemblies.
- Failure modes associated with interfaces ignored.

FMEA is a qualitative method in nature, and relies much on team experiences and opinions. To reduce the subjectivity, many researchers introduced quantitative analysis and methods, for example, defining the ratings of severity, occurrence, and detection based on actual data introduces quantitative aspects to FMEA (Xu et al. 2020). Popular research topics include tests (Frosini 2016), fuzzy data analysis (Balaraju et al. 2019), and computer simulation (Colman et al. 2019).

**Alternative Approaches**

Quality professionals have proposed a few alternative processes sharing the predictive and preventive nature of FMEA. In Chapter 4, DRBFM is discussed, which is about the same as FMEA in terms of principle and process. Although DRBFM is built for engineering design, one may adopt it for the applications in process and service development.

Another example is the following three-phase process for FMEA development (Ganot 2015). The three steps may help remove tedious team activities, and streamline the overall process (Kluse 2018).

1. Conduct a top-down development led by an FMEA facilitator and subject matter expert. The author argued that the common FMEA development was not effective in identification of event combinations, e.g. redundancies and interactions, for assessing total system reliability. In the top-down approach, an analysis begins by investigating unwanted systems failures to determine their related components.
2. Consider different weights for the three factors, for example, 100 for the severity (S), 10 for the occurrence probability (O), and 1 for the detectability (D). Then use a significant rank index (SRI = S + O + D) to replace the conventional RPN. Ganot (2015) suggested SRI with five severity levels and three occurrence levels. These are listed in Table 5.8, as a reference.
3. Use an action summary table to aggregate and categorize recommended actions.

**Table 5.8** Example of severity and occurrence in SRI

| Factor | Level | Description |
| --- | --- | --- |
| Severity (S) | Safety | May cause death or system destruction |
| | Catastrophic | May cause severe injury or system damage |
| | Critical | May cause minor injury or system shortcoming |
| | Major | May cause performance degradation |
| | Minor | May not influence performance but unscheduled work |
| Occurrence (O) | Probable | Several times in the life |
| | Remote | Unlikely but possible to occur |
| | Improbable | Assumed may not occur |

## 5.3.2   Applications of FMEA

As a proactive approach, FMEA has been widely applied in various fields. A few examples will be cited here to help readers better understand the principle. Please note that various formats are used across industries and disciplines, which may deviate from the common format shown in Table 5.1. In the published examples, there are some blanks in the FMEA forms.

### For Process and Product

FMEA has been widely used in manufacturing process planning and design. Table 5.9 shows one item (or row) in an FMEA for production processes (Lux et al. 2016). In this case, severity and occurrence probability for a failure did not change after the recommended action was taken. The detection of a failure was improved to $D_2$ from $D_1$. The updated $RPN_2$ was reduced to 30 from the original $RPN_1 = 75$.

Product design is another common application area of FMEA, called design FMEA. Table 5.10 gives an example of design FMEA jointly with other approaches for an in-vehicle information system (Li and Zhu 2020). As the potential problems are identified and evaluated, the following step (not shown in the original publication) is to find out the solutions needed to improve.

### For Services

More and more nonmanufacturing industries are using FMEA as a preventive measure. For example, in the pharmaceutical industry, FMEAs are used to identify the risk-based quality management system required by the various regulatory agencies, such as the International Council for Harmonisation (ICH). The ICH and FDA (Food and Drug Administration) have Q9 Quality Risk Management guidelines that suggest FMEA as a tool (ICH and FDA 2006). The food industry uses a variation of the FMEA process called HACCP (hazard analysis and critical control points) (FDA 2018).

An empirical study was conducted for an aging-in-place service using FMEA, based on 227 survey responses (Chen 2016). The FMEA was used to identify the improvement priorities among 16 service demands. The item in Table 5.11 was identified as the second priority for improvement. Table 5.12 shows another example of service FMEA, in which an RPN was reduced to 32 from 192 (McCain 2006).

### For Business Risk Analysis

Risk analysis and management are discussed in Chapter 2. As a proactive approach to predict and deal with business issues, FMEA is a suitable and effective method for risk management. Focusing on risks, instead of failure modes, a risk FMEA format needs minor modifications from the standard FMEA format. Table 5.13 shows a comparison between a common format (Table 5.1) and risk

**Table 5.9** Example of FMEA for manufacturing process

| 1. Item name and function | 2. Potential failure mode | 3. Potential failure effect | 4. Failure severity ($S_1$) | 5. Possible cause of failure | 6. Occurrence probability ($O_1$) | 7. Current control | 8. Failure detection ($D_1$) | 9. RPN$_1$ (=$S_1 \times O_1 \times D_1$) | 10. Recommended action | 11. Responsibility | 12. Action taken | 13. Post-action severity ($S_2$) | 14. Post-action occurrence ($O_2$) | 15. Post-action detection ($D_2$) | 16. Post-action RPN$_2$ |
|---|---|---|---|---|---|---|---|---|---|---|---|---|---|---|---|
| Clip fitting | No clip | Noisy product | 5 | Operator oversight | 3 | Visual inspection | 5 | 75 | Using passing sensor | AL | 2/14 | 5 | 3 | 2 | 30 |

**Table 5.10** Example of FMEA for product design

| 1. Item Name and Function | 2. Potential Failure Mode | 3. Potential Failure Effect | 4. Failure severity ($S_1$) | 5. Possible Cause of Failure | 6. Occurrence Probability ($O_1$) | 7. Current Control | 8. Failure Detection ($D_1$) | 9. RPN$_1$ (=$S_1 \times O_1 \times D_1$) | 10. Recommended Action | 11. Responsibility | 12. Action Taken | 13. Post-action severity ($S_2$) | 14. Post-action occurrence ($O_2$) | 15. Post-action Detection ($D_2$) | 16. Post-action RPN$_2$ |
|---|---|---|---|---|---|---|---|---|---|---|---|---|---|---|---|
| Category A9, Step 1.1.1 | User misunderstanding the meaning of icons | Influence on subsequent operations | 7 | Meaning of icon is not clear | 8 | | 4 | 224 | | | | | | | |

**Table 5.11** Example of FMEA for service

| 1. Item name and function | 2. Potential failure mode | 3. Potential failure effect | 4. Failure severity $(S_1)$ | 5. Possible cause of failure | 6. Occurrence probability $(O_1)$ | 7. Current control | 8. Failure detection $(D_1)$ | 9. $RPN_1$ $(=S_1 \times O_1 \times D_1)$ | 10. Recommended action | 11. Responsibility | 12. Action taken | 13. Post-action severity $(S_2)$ | 14. Post-action occurrence $(O_2)$ | 15. Post-action detection $(D_2)$ | 16. Post-action $RPN_2$ |
|---|---|---|---|---|---|---|---|---|---|---|---|---|---|---|---|
| Friends to be get along with well | Customer importance 4.7 | Not meeting (empathy) demand | 9 | Transportation and facilities recreational activities | 9 | | 10 | 810 | | | | | | | |

**Table 5.12** Another example of FMEA for service

| 1. Item name and function | 2. Potential failure mode | 3. Potential failure effect | 4. Failure severity $(S_1)$ | 5. Possible cause of failure | 6. Occurrence probability $(O_1)$ | 7. Current control | 8. Failure detection $(D_1)$ | 9. $RPN_1$ $(=S_1 \times O_1 \times D_1)$ | 10. Recommended action | 11. Responsibility | 12. Action taken | 13. Post-action severity $(S_2)$ | 14. Post-action occurrence $(O_2)$ | 15. Post-action detection $(D_2)$ | 16. Post-action $RPN_2$ |
|---|---|---|---|---|---|---|---|---|---|---|---|---|---|---|---|
| Implement and manage on-site recruiting process | No on-site recruiting | Employees not available | 8 | Not available for on-site | 8 | | 3 | 192 | Cross training all recruiter on the process | Branch managers | Recruiters cross trained | 8 | 2 | 2 | 32 |

**Table 5.13** FMEA format for risk analysis

| Type | Step 1 | Step 2 | | | | | Step 3 | | |
|---|---|---|---|---|---|---|---|---|---|
| Standard FMEA | 1. Item name and function | 2. Potential failure mode | 3. Potential failure effect | 4. Failure severity $(S_1)$ | 5. Possible cause of failure | 6. Occurrence probability $(O_1)$ | 7. Current control | 8. Failure detection $(D_1)$ | 9. $RPN_1 = (=S_1 \times O_1 \times D_1)$ |
| Risk FMEA | 1. Item name and func-tion | 2. Risk form | 3. *Risk* effect | 4. *Impact* severity $(S_1)$ | 5. Possible cause of *risk* | 6. Occur-rence proba-bility $(O_1)$ | 7. Current control | 8. *Risk* detection $(D_1)$ | 9. $RPN_1 = (=S_1 \times O_1 \times D_1)$ |

**Table 5.14** Example of impact rating for risk FMEA

| Rating | Schedule | Cost | Technical |
|---|---|---|---|
| 9–10 | Major milestone impact and > 20% impact to critical path | Total project cost increase > 20% | The effect on the scope renders end item unusable |
| 7–8 | Major milestone impact and 10–20% impact to critical path | Total project cost increase of 10–20% | The effect on the scope changes the output and it may not be usable to client |
| 5–6 | Impact of 5–10% impact to critical path | Total project cost increase of 5–10% | The effect on the scope changes the output and requires client approval |
| 3–4 | Impact of < 5% impact to critical path | Total project cost increase of < 5% | The effect on the scope is minor, but requires an approved scope change |
| 1–2 | Impact insignificant | Project cost increase insignificant | Changes are not noticeable |

FMEA format. As already discussed, a risk score is the product of the impact severity (S) and occurrence probability (O) for a specified risk. An RPN, discussed earlier, may be used separately from a conventional RPN, or combined with the RPN for a treatment prioritization.

Recall that the ratings of parameters – i.e. impact severity (S), occurrence probability (O), and risk detection (D) – can be case dependent. One study on project risk FMEA also suggested schedule, cost, and technical factors of a project, as shown in Table 5.14 (Carbone and Tippett 2004).

## 5.4 Control Plan

### 5.4.1 Basics of Control Plan

#### Process of Control Plan

A control plan is not a general term here but a quality planning and control approach and document that are often used in new product, process, and service development. From its functions, a control plan can be viewed as:

- A small project for quality improvement, with a defined schedule and accountable party.
- A summary description of work, and checklist, in a development or operation phase.
- A tracking document of the quality issues of a product or service, along with the work being conducted and updated status.
- A work guideline, monitoring major problem-solving activities, using checkpoints to assure meeting requirements prior to passing along to the next operation.

The purpose of using control plans is to validate accuracy or correctness of an operation before moving forward within a given timeframe. A control plan can be used during different phases, such as prototyping, pre-launch trials, operations (production or routine business operations), and with specific focuses on a function or feature. For a family of a project or service with very similar features, a general control plan can be developed for the family. For continuous improvement, a control plan can be carried out during normal production or operation.

As an effective tool for problem solving and action tracking, a control plan is often required in supplier quality management. For example, in the production part approval process (or PPAP, discussed in-depth in Chapter 6), a control plan is an integral element.

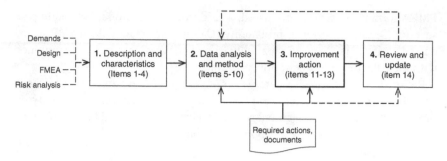

**Figure 5.13**  Overall process flow of control plan development.

For the automotive sector, the international standard IATF 16949:2016 and the advanced product quality planning (APQP) (AIAG 2008) standard suggest having a control plan, itself in a standard format. Similar to FMEA, the overall process of control plan development and documentation are shown in Figure 5.13, where FMEA can be an input and driver to a control plan. From this point, a control plan is considered as a follow-up action plan to the FMEA recommendations. Thus, the fundamental value of a control plan is its functions following FMEA recommended actions.

**Format of Control Plan**

As a structured approach, a control plan has formalized items and formats for quality issues and actions, with a general format shown in Table 5.15. AIAG has a template and example for product development in the automotive industry (AIAG 2008).

There are 14 items in a control plan that can be grouped into four technical sections. Thus, the control plan development is broken into four steps.

The <u>first step</u> is to identify the description and characteristics of the items needed for a control plan:

1. Process/product/function. This is a brief description of the items to be controlled.
2. Characteristics to quality. The characteristics that are identified as critical or special/significant to customer expectations should be included in a control plan. The details of these characteristics can be obtained from other work, such as design (discussed in Chapters 3 and 4).
3. Technical specifications. They are the design specifications, such as a dimensional tolerance and required time limits, of the characteristics of a product, process, or service.
4. Equipment, tooling, etc. This column lists the main hardware, such as machinery, facilities, and tooling, related to the characteristics in item 2.

**Table 5.15** Common format of a control plan

| Part #: | Name: | | Original date: | CP #: |
|---|---|---|---|---|
| Owner: | Org.: | | Revision date: | Status: |
| Team: | | | Approval date: | |

| 1) Description and characteristics | | | | 2) Data collection, analysis, and method | | | | | | 3) Action | | | 4) Update |
|---|---|---|---|---|---|---|---|---|---|---|---|---|---|
| 1. Process/product/function | 2. Characteristic to quality | 3. Technical specifications | 4. Equipment, tooling, etc. | 5. Description of characteristics | 6. Evaluation/measurement | 7. Sample size and conditions | 8. Measurement frequency | 9. Measurement locations | 10. Measurement operators | 11. Control action taken | 12. Action lead/team | 13. Applicable standards | 14. Evaluation review |

The <u>second step</u> is the notes on data collection and analysis, as well as methods used in a control plan. The information in this step is 1) Below is a numbered list. Refer to the manuscript.2) The format of the list should be the same as the previous one.control-action related:

5. Description of characteristics for data collection and analysis.
6. Evaluation/measurement method to be used for data analysis.
7. Sample size and condition of the data collection.
8. Measurement frequency, e.g. collecting data every two hours.
9. Measurement locations.
10. Names of measurement operators.

The <u>third step</u> is the records of the identification, control, and improvement actions to mitigate failures and/or prevent them from being passed to the next step:

11. Description, plan, and date of the control actions taken.
12. Names of lead or team members for the action taken.
13. Specific standards associated with the measures and actions, if applicable.

The <u>last step</u> is about updates and re-evaluation:

14. Evaluation reviews for updates and/or conclusion.

Depending on disciplines and applications, control plans have much more variation of format than FMEA, as seen with numerous examples shown in Tables 5.17 and 5.18 in section 5.4.3.

### 5.4.2 Considerations in Control Plan

**Control Plan Administration**

Discussed in Chapter 3, the characteristics of a product, process, and service can be categorized into three types: critical, special/significant, and standard/ordinary. Critical characteristics, such as important functions, dimensions, and types of materials, should be the focus of a control plan, as these characteristics have known impacts on safety, compliance, etc. From this point, a QFD and/or FMEA can help identify the critical characteristics for a given control plan. The special/significant characteristics of an effort are often included in a control plan because of their impacts on functions and quality.

To address these characteristics and manage a control plan, an overall summary sheet may be developed to list which characteristics should be covered in it. Table 5.16 illustrates three examples for product, process, and service, respectively. The development teams and management decide on the scope and coverage of a control plan.

**Table 5.16** Example of characteristics and control plan coverage

| Target | Characteristics | Critical | Special/ significant | Specification | Control plan |
|---|---|---|---|---|---|
| Product | Vehicle crash test | √ | | 4 star or better | Yes |
| | Cellphone battery life | | √ | 15 h standby | No |
| | ... ... | | | | |
| Process | Throughput | √ | | > 480 units/ shift | Yes |
| | Operation time | | √ | < 2 min + 10 s | Yes |
| | ... ... | | | | |
| Service | Prescription drug labels | √ | | > 99% | Yes |
| | Representative's knowledge | | √ | All questions answered | Maybe |
| | ... ... | | | | |

If a control plan is used for normal operations, it is a living document without a closure. One should keep updating such a control plan routinely, and use it as an operational reference and continuous improvement guide.

### Supporting Information

Sometimes a control plan needs unique information from other quality efforts and documents to support it, such as data analysis and control results. A control plan may also include a flow chart (ASQ 2016) and other types of graphics. For manufacturing productions and service operations, a process flow diagram, discussed in Chapter 8, shows an overall flow and the relationship between major equipment and operating personnel. The diagram's graphic information is helpful to control plan development.

The methods often used to support a control plan include:

- QFD (Chapter 3)
- Design reviews (Chapter 4)
- Prototyping (Chapter 4)
- 14 quality tools (Chapter 8)
- Problem solving (e.g. PDCA and 8D, Chapter 8)
- Design of Experiment (more technical, refer to dedicated books)

A control plan may be designed as a one-pager including multiline items, each describing a characteristic. This design makes it easy to review, follow up, and close out. As a work summary, a control plan only shows brief information on data collection and analysis. Supporting data-analysis and problem-solving documents can be an appendix to a control plan.

A control plan is neither an operational instruction nor a routine monitoring plan. After a control plan is successfully completed, its output can be a foundation for updating an operation instruction, standards, and/or a monitoring plan for routine operations. Many times, updated monitoring is further developed into a business system based on control plans.

### 5.4.3 Applications of Control Plan

The principle and development process behind a control plan show it can be used in various disciplines. A few illustrative examples will be given here for reference. As mentioned, different formats and focuses are common in control plan applications.

For a manufacturing process, a control plan was developed for incoming materials. A partial control plan is shown in Table 5.17 (Anjoran 2020).

In clinical laboratories, quality control plans must be reviewed and re-approved annually (Bruno 2019). The prerequisite of these control plans is conducting a risk analysis with five components: specimen, environment, testing personnel, reagents, and test system. From a risk analysis perspective, quality control plans have been developed on these details, including the number and the composition of specimens, test frequency, and the criteria for acceptability of specimens.

Control plans may be improvement projects in various operations. One particular control plan was developed for a project to improve mail handling and rerouting (Kemper et al. 2011). In this case, the control plan defined the functions and responsibility of the handling process and specified control limits, etc. (see Table 5.18). The control plan was an overall guide linked with the OCAP (out-of-control action plan). This OCAP provided detailed work instructions for quality issues.

In another application, a control plan was used to identify the maintenance needs for roadway bridge management. The authors proposed a standardization for European highway bridges, as shown in Figure 5.14, which combined a process flow and control plan document (Casas and Matos 2017). The quality control plan was tied to performance goals, e.g. traveling time, traffic allowance, safety level, and serviceability of different types of bridges.

**Table 5.17** Example of a process control plan.

| 1. Process number | 2. Work description | 3. Machine | 4. Number | 5. Product characteristics | 6. Process characteristics | 7. Special characteristics | 8. Spec/ tolerance | 9. Measurement | 10. Sample quantity | 11. Sampling frequency | 12. Control method | 13. Reaction plan |
|---|---|---|---|---|---|---|---|---|---|---|---|---|
| 20 | Fiber drawing/cut | Subcontract | 11 | Diameter | | B | $\Phi 31.8_{-0.05}$ | Caliper | 2 | Per lot | Sampling | Return to supplier |

**Table 5.18** Example of a process improvement control plan.

| 1. Measurement | 2. Who | 3. How | 4. Where | 5. When | 6. Reporting | 7. Norm/ spec | 8. Lower limit | 9. Which OCAP |
|---|---|---|---|---|---|---|---|---|
| Correct sticker use | Regional manager | Sampling with measurement form | Output steam of each team | Monthly | Sr. dept. controller | 95% | 85% | HPB2 |
| Machine processed rerouted mail | Sr. dept. controller | System data | Input stream of sorting machines | Monthly | Manager sorting hub | 85% | 80% | HPS2 |

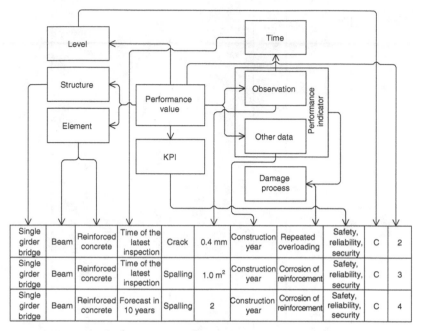

**Figure 5.14** Example of control plan development. *Source*: Casas, J.C. and Matos, J.C. (2017). Standardization of quality control plans for highway bridges in Europe: COST Action TU 1406, IOP Conference Series: Materials Science and Engineering, 236, 012051. 10.1088/1757-899X/236/1/012051.

## Summary

### Understanding FMEA

1 The objectives of doing an FMEA are to identify and evaluate potential failure modes, propose preventive actions, and follow up the improvement plans.

2 Depending on targets, FMEA can be on designs, processes, services, risks, etc.

3 Inputs to an FMEA include specifications, standards, predefined criteria, pre-analyses, historical data, etc.

4 Typical outputs of an FMEA are improvement recommendations or an AP list.

5 FMEA development normally involves five steps to address 16 items on an FMEA form.

6 Parameters and rating criteria in FMEA are application dependent, with industry-specific guidelines.

### Pre- and Post-work of FMEA

**7** Boundary diagrams, interface matrixes, and/or parameter diagrams are often conducted as pre-FMEA analysis.

**8** The fundamental criteria for FMEA outputs are based on the ratings for severity, occurrence, and detection.

**9** An RPN $(= S \times O \times D)$ is a common criterion, while an AP list is a relatively new criterion.

**10** There are three levels of recommendation based on either RPN or AP. Individual parameters should be considered as well.

**11** Follow-up and management reviews help ensure FMEA completion.

### Implementation of FMEA

**12** Decision of ratings for severity, occurrence, and detection should be data driven, as much as possible.

**13** Having good teamwork and management support are necessary to an FMEA's success.

**14** FMEA often serves as a living technical document, for routine updates and reviews.

**15** There are similar quality approaches to FMEA, e.g. DRBFM.

**16** There are significant variations for FMEA applications, in terms of ratings, criteria, processes, etc.

### Control Plan

**17** A control plan is a quality-planning and control approach, and considered a monitoring and verification document.

**18** A control plan is normally required for the critical characteristics of a product, process, or service; it is also recommended for special/significant characteristics.

**19** A control plan can be supported by other quality approaches, e.g. QFD, PDCA, etc.

**20** Control plans vary extensively for various applications.

**21** A control plan can support updating an operation instruction, standards, and/or be a monitoring plan for routine operations.

## Exercises

### Review Questions

**1** Explain the objectives of FMEA with an example.

**2** Discuss the input information and requirements for FMEA.

3  Discuss the outputs and their values for FMEA.

4  Explain the benefits of FMEA applications with examples.

5  Describe the steps of FMEA development.

6  During FMEA development, a team has different options on ratings. One member suggests voting for them. Do you agree? Why?

7  Discuss the RPN criteria of FMEA for follow-up actions.

8  Discuss possible changes in S rating after follow-up actions.

9  Explain the roles of a boundary diagram, interface matrix, or parameter diagram for FMEA development.

10  "Different weights should be used for S, O, and D in an RPN calculation." Do you agree with this statement? Why or why not?

11  Discuss the necessity of considering O and D for recommendations, in case where S and RPN are low.

12  One says that FMEA is a living document. Another says that FMEA should be like a project with a lifespan and closure time. Provide your viewpoint and justification.

13  Identify some differences between a design FMEA and a service FMEA.

14  Compare a design FMEA and a risk FMEA for their differences.

15  Discuss the decision follow-up flow of FMEA recommendations.

16  Use examples to show the input to or driving force of a control plan.

17  Discuss the necessity of a control plan that is based on the characteristics of a product or service.

18  Discuss the considerations in control plan applications.

19  Compare two control plans in the service sector.

20  Use examples to show how another quality tool can support a control plan.

## Mini-project Topics

1  Develop a simple FMEA on your new product, process, or service, which is still in a development phase or to be developed (not in operation or in place yet).

2  Review the effectiveness of FMEAs based on published applications or your observation/experience.

3  How would you develop a scale of 1–10 for the severity rating in your own professional field?

4  Find an example of S, O, or D rating of FMEA in a professional field, and discuss the rating.

5  Review an application of using a boundary diagram for FMEA.

6  Review an application of using an interface matrix for FMEA.

7  One states that S, O, and D ratings, RPNs, and AP are subjective, which makes FMEA recommendations less robust. Assess the statement and provide your viewpoint.

8  Search and find five FMEA applications and discuss their RPN criteria.
9  Use library databases, or https://scholar.google.com, to find new developments of FMEAs or control plans.
10  Review the commonality between a control plan and a problem-solving tool (e.g. 8D) in Chapter 8.

## References

AIAG. (2008). *Advanced Product Quality Planning and Control Plan*, 2nd edition. Southfield, MI: Automotive Industry Action Group.

AIAG and VDA. (2019). *AIAG & VDA FMEA Handbook*. Southfield, MI: Automotive Industry Action Group (AIAG) and German Association of the Automotive Industry (VDA).

Aljazzar, H., Fischer, M., Grunske, L. et al. (2009). Safety analysis of an airbag system using probabilistic fmea and probabilistic counterexamples, *2009 Sixth International Conference on the Quantitative Evaluation of Systems*, September 13–16, 2009, Budapest, Hungary, pp. 299–308. 10.1109/QEST.2009.8.

Anjoran, R. (2020). How to fill out a process control plan to raise product quality, January 9, 2020, https://www.cmc-consultants.com/blog/how-to-fill-out-a-process-control-plan-to-raise-product-quality, accessed in May 2020.

ASQ. (2016). Control plan, service quality division, http://asqservicequality.org/glossary/control-plan, accessed in April 2020.

Balaraju, J., Raj, M.G., and Murthy, C.S. (2019). Fuzzy-FMEA risk evaluation approach for LHD machine – A case study, *Journal of Sustainable Mining*, Vol. 18, Iss. 4, pp. 257–268. 10.1016/j.jsm.2019.08.002.

Bruno, L.C. (2019). Individualized quality control plan: 3 years later, *Clinical Microbiology Newsletter*, Vol. 41, Iss. 12, pp. 103–109. 10.1016/j.clinmicnews.2019.05.003.

Carbone, T.A. and Tippett, D.D. (2004). Project risk management using the project risk FMEA, *Engineering Management Journal*, Vol. 16, Iss. 4, pp. 28–35. 10.1080/10429247.2004.11415263.

Casas, J.C. and Matos, J.C. (2017). Standardization of quality control plans for highway bridges in Europe: COST Action TU 1406, *IOP Conference Series: Materials Science and Engineering*, 236, 012051. 10.1088/1757-899X/236/1/012051.

Cassanelli, C., Mura, G., Fantini, F. et al. (2006). Failure analysis-assisted FMEA, *Microelectronics Reliability*, Vol. 46, pp. 1795–1799. 10.1016/j.microrel.2006.07.072.

Catton, W.A., Walker, S.B., McInnis, P. et al. (2019). Design and analysis of the use of Re-purposed electric vehicle batteries for stationary energy storage in Canada, *Batteries*, Vol. 5, Iss. 14, p. 19. 10.3390/batteries5010014.

Chen, S. (2016). Determining the service demands of an aging population by integrating QFD and FMEA method, *Quality and Quantity*, Vol. 50, Iss. 1, pp. 283–298. 10.1007/s11135-014-0148-y.

Chuang, P. (2007). Combining service blueprint and FMEA for service design, *The Service Industries Journal*, Vol. 27, Iss. 2, pp. 91–104. 10.1080/02642060601122587.

Colman, N., Stone, K., Arnold, J., Doughty, C. et al. (2019). Prevent safety threats in new construction through integration of simulation and FMEA, *Pediatric Quality & Safety*, Vol. 4, Iss. 4, 10.1097/pq9.0000000000000189.

Enoch, O.F., Shuaib, A.A., and Hasbullah, A.H. (2015). Applying P-diagram in product development process: an approach towards design for six sigma, *Applied Mechanics and Materials*, Vols. 789–790, pp. 1187–1191. 10.4028/www.scientific. net/AMM.789-790.1187.

FAA. (2011). AC 23.1309-1E - system safety analysis and assessment for part 23 airplanes, federal aviation administration, https://www.faa.gov/regulations_policies/advisory_circulars/index.cfm/go/document.information/documentID/1019681, accessed in October 2020.

FDA. (2018). Hazard analysis critical control point (HACCP), Content current as of: 01/ 29/2018, https://www.fda.gov/food/guidance-regulation-food-and-dietary-supplements/hazard-analysis-critical-control-point-haccp, accessed in September 2020.

Frosini, F., Miniati, R., Grillone, S. et al. (2016). Integrated HTA-FMEA/FMECA methodology for the evaluation of robotic system in urology and general surgery, *Technology & Health Care*, Vol. 24, Iss. 6, pp. 873–887. 10.3233/THC-161236.

Ganot, A. (2015). Lean (and Friendly FMEA), *Presentation to International Applied Reliability Symposium*, Tucson AZ, June 4, 2015. https://www.isq.org.il/wp-content/uploads/Quality_2015_lectures_2.4_2_Ganot_PP.pdf, accessed in August 2020.

IATF. (2016). IATF 16949 Quality management system requirements for automotive production and relevant service parts organizations, International Automotive Task Force, https://www.iatfglobaloversight.org.

ICH and FDA. (2006). Q9 quality risk management, FDA-2005-D-0334, https://www.fda.gov/regulatory-information/search-fda-guidance-documents/q9-quality-risk-managementandhttps://database.ich.org/sites/default/files/Q9%20Guideline.pdf, accessed in September 2020.

IEC. (2018). *IEC 60812:2018 Failure Modes and Effects Analysis (FMEA and FMECA)*, Geneva, Switzerland: International Electrotechnical Commission.

ISO. (2017). *ISO 12132:2017 Plain Bearings – Quality Assurance of Thin-Walled Half Bearings – Design FMEA*, Geneva, Switzerland: International Organization for Standardization.

James, F. (2009). Toyota recalls 3.8 Million cars for floor mats linked to stuck gas pedals, NPR, September 29, 2009. https://www.npr.org/sections/thetwo-way/2009/09/toyota_recalls_38_million_cars.html, accessed in August 2020.

Johnson, K.G. and Khan, M.K. (2003). A study into the use of the process failure mode and effects analysis in the automotive industry in the UK, *Journal of Materials Processing Technology*, Vol. 139, Iss. 1–3, pp. 348–356. 10.1016/S0924-0136(03)00542-9.

Kemper, B., Koning, S., Luijben, T., and Does, R. (2011). Quality Quandaries: Cost and Quality in Postal Service, *Quality Engineering*, Vol. 23, Iss. 3, pp. 302–308. 10.1080/08982112.2011.575754.

Kluse, C. (2018). The 25th anniversary of the AIAG FMEA reference manual: A systematic literature review of alternative FMEA methods, *Journal of Management and Engineering Integration*, Vol. 11, Iss. 21, pp. 38–46.

Kluse, C. (2020). A critical analysis of the AIAG-VDA FMEA; Does the newly released AIAG-VDA method offer improvements over the former AIAG method?, *Journal of Management and Engineering Integration*, Vol. 13, Iss. 1, pp. 71–85.

Li, Y. and Zhu, L. (2020). Risk analysis of human error in interaction design by using a hybrid approach based on FMEA, SHERPA, and fuzzy TOPSIS, *Quality and Reliability Engineering International*, Vol. 36, Iss. 5, pp. 1657–1677. 10.1002/qre.2652.

Lux, A., Bikond, J., Etienne, A., and Quillerou-Grivot, E. (2016). FMEA and consideration of actual work situations for safer design of production systems, *International Journal of Occupational Safety and Ergonomics*, Vol. 22, Iss. 4, pp. 557–564. 10.1080/10803548.2016.1180856.

McCain, C. (2006). Using an FMEA in a service setting, *Quality Progress*, September 2006, pp. 24–29.

NASA. (2010). Standard for performing a failure mode and effects analysis (FMEA) and establishing a critical items list (CIL), US National Aeronautics and Space Administration, https://rsdo.gsfc.nasa.gov/documents/Rapid-III-Documents/MAR-Reference/GSFC-FAP-322-208-FMEA-Draft.pdf, accessed in April 2020.

NBC. (2006). Ford recolling nearly 20,000 Mustang cobras – gas pedal on sports cars can catch in floor carpeting, lead to crash, April 11, 2006. http://www.nbcnews.com/id/12270265/ns/business-autos/t/ford-recalling-nearly-mustang-cobras/#.X0Kws8hKiHc, accessed in August 2020.

Rastayesh, S., Bahrebar, S., Blaabjerg, F. et al. (2020). A system engineering approach using FMEA and Bayesian network for risk analysis – A case study, *Sustainability*, Vol. 12, Iss. 1, 18 pages. 10.3390/su12010077.

SAE. (1996). *Guidelines And Methods For Conducting The Safety Assessment Process On Civil Airborne Systems And Equipment*, ARP4761, Warrendale, PA: SAE International.

SAE. (2009). *Potential Failure Mode And Effects Analysis In Design (Design FMEA), Potential Failure Mode And Effects Analysis In Manufacturing And Assembly Processes (Process FMEA), J1739*. Warrendale, PA: SAE International.

SAE. (2012). *Recommended Failure Modes and Effects Analysis (FMEA) Practices for Non-Automobile Applications*, ARP5580, Warrendale, PA: SAE International.

Spreafico, C., Russo, D., and Rizzi, C. (2017). A state-of-the-art review of FMEA/FMECA including patents, *Computer Science Review*, Vol. 25, pp. 19–28. 10.1016/j.cosrev.2017.05.002.

Tang, H. (2018). *Manufacturing System and Process Development for Vehicle Assembly*. Warrendale, PA: SAE International.

Tanka, T., Moriya, D., and Yamaur, T. (2014). The quality analysis of the video game failure, *Proceedings of the International Conference on Computer Science, Computer Engineering, and Education Technologies*, pp. 26–39. Kuala Lumpur, Malaysia, 2014.

Winter, D., Ashton-Rickardt, P., Ward, C. et al. (2015). An enhanced risk reduction methodology for complex problem resolution in high value, low volume manufacturing scenarios, *SAE International Journal Materials and Manufacturing*, Vol. 9, Iss. 1, pp. 49–64. 10.4271/2015-01-2595.

Wolniak, R. (2019). Problems of use of FMEA method in industrial enterprise, *Production Engineering Archives*, Vol. 23, pp. 12–17. 10.30657/pea.2019.23.02.

Xu, Z., Dang, Y., Munro, P. et al. (2020). A data-driven approach for constructing the component-failure mode matrix for FMEA, *Journal of Intelligent Manufacturing*, Vol. 31, pp. 249–265. 10.1007/s10845-019-01466-z.

# 6

# Supplier Quality Management and Production Part Approval Process

## 6.1 Introduction to Supplier Quality

### 6.1.1 Supplier Quality Overview

**Significance of Suppliers**

In today's world, few product producers or service providers can do all the work for their customers because of various reasons, such as business focus, financial constraint, and technical expertise. An organization often outsources some supportive functions, and even a small segment of core functions, to its suppliers.

For a manufacturer, a supplier may provide raw materials, production parts, and service parts. A supplier may also handle some specialized manufacturing processes of products, such as casting, heat-treating, and surface finishing. On the service side, a common example is an IT function that can be outsourced to a professional provider. All the outsourced functions and suppliers play an integral role to the product or service quality experienced by customers, as discussed in earlier chapters.

Depending on the significance of suppliers' work and their deliverables to a producer or provider, their roles may be categorized into three levels (see Figure 6.1). Their significance levels can be determined based on functionality, operation, quality, etc. by a producer or provider.

1. Essential: These suppliers are the crucial part of customers' business operations. One example would be a supplier that provides production parts to a product producer. Without the supplier's parts, the producer would shut down their manufacturing operations. In terms of quality, the essential suppliers' work directly affects, or even plays a determinative role, in a producer or provider's quality for their customers. Many product recalls are related to supplier quality issues. For example, one might investigate

*Quality Planning and Assurance: Principles, Approaches, and Methods for Product and Service Development*, First Edition. Herman Tang.

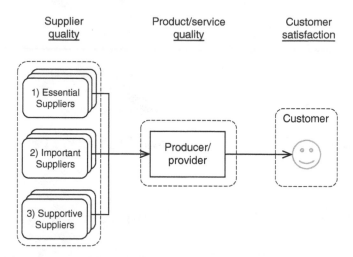

**Figure 6.1**  Roles of suppliers to producer/provider and their ultimate customers.

the comparable tire quality between Ford and Firestone (NHTSA 2001). It is generally necessary to identify and qualify backups for essential suppliers. In some cases, their functionality is irreplaceable, or very difficult to substitute.

2. Important: Such suppliers may not be necessary for minute-to-minute operations of a producer or provider. A customer-centric company may keep its operations temporal, without such a supplier's work or support. These suppliers may be replaced, but with adverse effects, including low effectiveness or delays in operations. IT support for non-core business operations may belong to this category, while the IT functions and support for core operations should be considered essential.

3. Supportive: These suppliers assist in optimizing the operations of a producer or provider. A producer or provider may keep running with limited or no such assistance provided by these suppliers. Examples here include dining and janitorial services. Supportive functions have indirect impacts on operations and quality.

The purpose of identifying the significant roles of suppliers is for strategic thinking and planning. The more important a supplier is to a business, the more management pays attention and works closely with these suppliers for their quality assurance. When evaluating suppliers, remember that there is ambiguity, and overlap between the three categories. For example, a supplier can be both important and supportive. A supplier might be rated as 2.3 on a scale of 1–3, where 1 is of the highest necessity.

**Supplier Quality Management**

Supplier quality is a supplier's ability to deliver customer goods or services. The overall requirements for supplier quality are:

- Meeting current expectations consistently
- Ability for continuous improvement, toward meeting future requirements

Supplier quality management (SQM) is a broad term, which covers a life cycle from pre-contract to post-contract activities, including strategic sourcing, supplier development, supplier qualification, conflict resolution, and purchase ordering (PO) process. From a standpoint of quality management, there are four major components (see Figure 6.2). The fourth component is not a focus of this book, so it is discussed briefly.

1. Supplier quantity planning
2. Sourcing/supplier selection
3. Supplier quality assurance
4. Supplier quality monitoring and improvement

SQM should be a system of partnership, in principle, approach, process, and inter-department effort. It is in the best interest of a company to select and work with suppliers who can ensure high quality of a product or service for the company's ultimate customers. A study based on 238 plants in the US, Japan, Italy, Sweden, Austria, Korea, Finland, and Germany showed that the internal quality

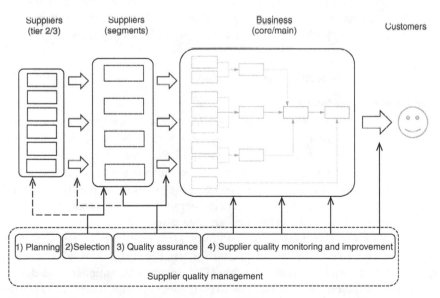

**Figure 6.2** Sourcing and quality assurance in SQM.

management of customer companies had a significantly positive impact on the SQM with a standardized coefficient of 0.62 (Zeng et al. 2013).

SQM should focus not only on supplier outputs but also on supplier's processes to ensure their outputs. Concentrating on a supplier's processes, supplier management is more proactive and preventive than reactive. A large amount of crisis response over supplier quality implies room for improvement on a supplier's processes.

To be proactive, potential risks and issues should be addressed in early supplier selection and quality assurance phases. Five major risks were identified for an external supply chain, including:

"● Demand risks – caused by unpredictable or misunderstood customer or end-customer demand
● Supply risks – caused by any interruptions to the flow of product, whether raw material or parts, within your supply chain
● Environmental risks – from outside the supply chain; usually related to economic, social, governmental, and climate factors, including the threat of terrorism
● Business risks – caused by factors such as a supplier's financial or management stability, or purchase and sale of supplier companies
● Physical plant risks – caused by the condition of a supplier's physical facility and regulatory compliance." (Queensland 2020)

This chapter mainly discusses the quality planning related issues of SQM on critical suppliers. The same principles and procedures can apply to important and supportive suppliers when resources are available and justified.

### 6.1.2 Supplier Selection and Evaluation

#### Process of Supplier Selection

Sourcing planning and selection are essential in quality management. The fundamental goal here is to ensure suppliers work well with and are coordinated and integrated into future business, to serve ultimate customers. Thus, the primary objective for sourcing planning and selection is to select suppliers based on quality and other business requirements. A process flow chart of sourcing selection is shown in Figure 6.3 and can apply to a new business opportunity or a renewal cycle in an existing business arrangement.

A sourcing selection process starts with the identification of several supplier candidates for a major project or function. Then, the customer sends the supplier candidates an invitation to tender (ITT). The selected supplier candidates should make their tenders (or bids) based on the specific requirements described in this ITT. One common practice is "free-expression competitive bidding." In

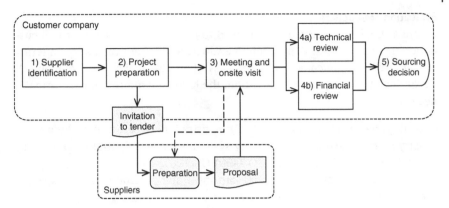

**Figure 6.3** Process flow of sourcing selection.

such a case, requirements can be general; the supplier candidates must prepare their detailed responses based on their individual capacity, strengths, and conditions. Most times, a customer and its supplier have a nondisclosure agreement (NDA) or confidentiality agreement, which reduces the risk of conflict of interests and protects the information assurance for both parties.

Normally, the company reads the suppliers' proposals, and conducts on-site assessments at suppliers' sites. Then, the company performs two types of internal reviews: of technical and financial aspects. The technical (e.g. on quality, capacity, delivery, facilities, risk, work force, etc.) and financial (cost-effectiveness, stability, credit, etc.) reviews are often conducted separately, by their respective departments, for confidentiality and to reduce conflict of interest.

In a technical review, one may assume costs are not an issue, and decide to choose supplier X over supplier Y considering the quality of the product, reliability of the service, and on-time delivery capability. Meanwhile, another team assesses the cost, assuming the suppliers can meet the minimum technical requirements. The financial review focuses on the financial feasibility and cost effectiveness of these supplier candidates.

Such evaluations, site visits, and internal reviews often require several iterations. During the reviews, supplier candidates on the final list may be summoned for clarification, additional information, and/or revision to existing information. The output of technical and financial reviews is a list of ranked supplier candidates for senior management decision.

It is possible that a supplier quotes their business at a very low price to win a trade agreement. In many cases, going for the suppliers with the lowest upfront costs leads to the higher risks of quality issues and additional costs later on, harming a business in the long term. Thus, being cautious is important in the comprehensive sourcing decision review, when considering the low bids. In practice, the lowest bidder is not always the one selected.

**Evaluation of Suppliers**

Before a supplier is selected and allowed to provide parts or services to a customer, there are several steps taken to evaluate a supplier's capability, in terms of performance (outcomes) and process (how to achieve the outcomes). Regarding quality aspects of evaluation, the following information is normally needed:

- Quality management system (QMS): registration or certification by a third party to an industrial sector quality system, such as ISO 9001 (ISO 2015 a), IATF 16949 (IATF 2016), and ISO 14001 (ISO 2015 b).
- Quality requirement: the supplier can meet and possibility exceed specific requirements.
- Technical capability and ability to launch a new product and service.
- Personnel qualification: education, training, experience, and problem-solving practice (with examples).
- Test reliability: the test results of products meeting the minimum useful life expectations.

As a tentative business partner, a supplier may be expected to do more beyond build and produce in the future. In such cases, their aforementioned technical capabilities are important. Additional related factors may include:

- Delivery performance: on-time delivery rate, the quality and quantity of production capacity (tryout samples not sufficient).
- Cost competitiveness (total cost).
- Financial stability, status, and strength for future growth.
- Advantageous or unique technology in place.
- Operational capability, such as machinery and facilities.

Cost competitiveness, for example, is more than just a low initial price. Other factors, e.g. costs of transactions, change management, communication, and problem resolution, all affect the total cost for a given long run of trade.

A practical method of supplier selection is called an analytic hierarchy process (AHP). AHP is a structured way to analyze supplier candidates based on all major factors. Figure 6.4 shows an example of evaluating and selecting suppliers. In the figure, the factors associated with quality and other aspects of a business are grouped separately. The selection of factors and their weights is case dependent.

For each of the layers of factors for supplier candidates in an AHP model, the total of weights should be one or 100, referring to Chapter 3, for weight selection and normalization. Once all suppliers are evaluated based on these predefined factors, the score of each supplier candidate can be calculated, and then suppliers can be ranked for selection. The AHP applications on supplier selection

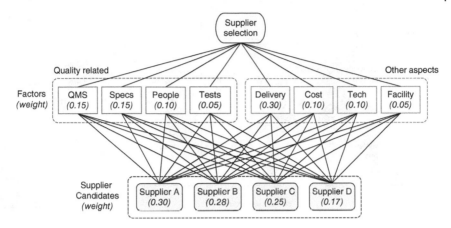

**Figure 6.4** Supplier evaluation using AHP method.

have been used in different industries, such as manufacturing (Hosseini and Khaled 2019), air transportation catering (Yan-Kai Fu 2019), and hospital pharmacy (Manivel and Ranganathan 2019).

Suppliers' overall technical and financial capabilities can also be compared in two-dimensional diagrams. Figure 6.5 (a) shows an overall assessment diagram of four separate suppliers. The ratings of these suppliers are based on an AHP analysis, and converted to a scale of 1–5, where 5 is superior. Using this comparison approach, the relative positions of supplier candidates are visualized. For example, supplier D falls into the conditional zone in the comparison diagram, due to its technical advantage balancing unfavorable financial aspects. Another supplier is rather balanced on financial and technical aspects, but is relatively low on both. Further evaluation reviews should be conducted on such suppliers. In addition, the evaluation can be more detailed for a supplier, based on its performance and process factors, for example, referring to Figure 6.5 (b). Further, these factors may have different weights in a given consideration.

### Supplier Quality Monitoring

Once a supplier is selected, the supplier's performance should be continuously monitored. Such an assessment is a baseline that the supplier may continue to cater toward, to gain additional business opportunities. For supplier quality monitoring, there are two main approaches, viewed as reactive and proactive.

1. Monitoring and follow-up. A customer constantly monitors supplier quality and follows up quality issues. Living monitoring documentation includes statistical process control (SPC) charts and quality metrics. Follow-up actions may include audits, specific investigations, and change requirements. There

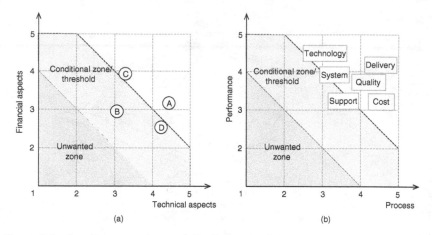

**Figure 6.5** Supplier comparison and detailed evaluation.

is a reaction and escalation process, with containment activities and actions. For example, the Food and Drug Administration (FDA) sent a warning letter to a drug manufacturer stating, "Until you correct all violations completely and we confirm your compliance with CGMP, the FDA may withhold approval of any new drug applications or supplements listing your firm as a drug manufacturer" (BioPham 2020).

2. <u>Certification program</u>. Certified suppliers can garner a high level of trust with a customer. For example, Ford has a well-known supplier certificate program, called Q1. To get a Q1 status, a Ford supplier must fulfill many requirements and expectations, demonstrating its excellence in a number of critical areas, such as:

- Capable systems: with third-party certifications in ISO/TS 16949 and ISO 14001, etc.
- Ongoing performance: no Ford intervention on supplier failure, 100% on-time delivery, consistently high-quality products, etc.
- Manufacturing site assessment: e.g. measurement system analysis (MSA) audit, evaluating readiness of the production process, quality system, requirements, evaluation methods, and/or corrective action plans.
- Customer satisfaction: zero quality complaints, no open corrective actions, etc., and directly endorsements from Ford's departments, e.g., manufacturing plants and material planning and logistics.
- Continuous improvement: the system, process, documentation, etc.

One routine task of SQM is to deal with various changes. The change management process is discussed in Chapter 7. Changes normally should be reviewed prior to execution, by going through a formal process if they are associated with noticeable risks. Common changes are:

- Process: technology, control, inspection, main parameters, automation, transportation, etc.
- Product: different sub-tier supplier, labeling, packaging, software, etc.
- Material: raw material, incoming material, component, etc.
- Facility: equipment relocation, facility-to-facility transfer, environmental conditions, etc.

For large and complex projects, supplier and source selection often compartmentalize to sub-suppliers, called tier-2 suppliers. Sometimes, a customer directs a tier-1 supplier in which tier-2 suppliers to use. In such a case, the customer and tier-1 supplier share the quality management activities and responsibilities for those tier-2 suppliers.

In the automotive industry, supplier performance is assessed following the production part approval process (PPAP), which is to be discussed in the following sections. Supplier assessment is often conducted prior to a PPAP agreement based on their quality capability, along with many non-quality aspects. Other industries and sectors have similar processes.

## 6.2 PPAP Standardized Guideline

### 6.2.1 Concept of PPAP

#### Principle of PPAP
When a supplier provides production parts as well as service parts for a product producer (i.e. a customer company), the supplier plays a critical role in the producer's business. As a mass production industry, the automotive industry realized the significant roles of production part suppliers and thus developed a guideline, called PPAP, in 1993 (AIAG 2006).

The purposes of a PPAP include:

- To ensure that the supplier is able to meet all the requirements of the parts, and supply these satisfactory parts to the customer.
- To provide evidence that the supplier (and its sub-tier suppliers) properly understood and fulfilled all customer requirements.
- To validate that the supplier is prepared for pre-launch volume to mass production outputs.
- To show that the supplier's process can produce parts that consistently meet requirements during an actual production run, at the quoted production rate.

Therefore, a PPAP is a business procedure in SQM (Figure 6.2). PPAP is specifically developed for automotive manufacturing, and applicable for other manufacturing industries, such as furniture and food industries and service sectors. There wil be more discussion on service sectors in later sections.

A PPAP is considered formal documentation that includes the outcomes of parts' quality, as well as administrative efforts undertaken. PPAPs can be also viewed as a quality "goalkeeper" to inherently ensure part quality and part requirements are met before a producer can use those parts in their production.

The PPAP principle and process can apply to relationships between tier-1 and tier-2 suppliers. If tier-2 suppliers are selected by a customer, the PPAP requirements for the tier-2 suppliers can be determined directly by the customer. For all scenarios, a PPAP remains a communication assurance in the network of customers and suppliers (see Figure 6.6).

The PPAP principle has been widely adopted by various industries beyond the automotive industry, including commercial equipment and appliances, and is considered a generic guideline for all types of suppliers and their customers. For example, the aerospace industry uses AS9145 as its PPAP standard (SAE 2016). Some companies require a PPAP as a prerequisite to complete a PO, last payments, or other types of contractual documents. Ireland (2017) carried out a study on the PPAP applications for military weapon systems production.

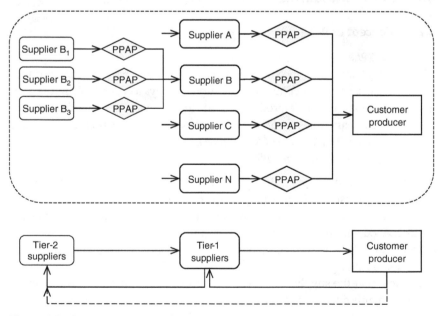

**Figure 6.6** Functionality of PPAP in supplier chain.

**Process of PPAP**

There are several review and approval steps in a PPAP procedure. The overall process and responsibilities of a customer and its suppliers are illustrated in Figure 6.7. In addition to their own responsibilities in the process, a customer and its suppliers as partners collectively work together toward the success of PPAP.

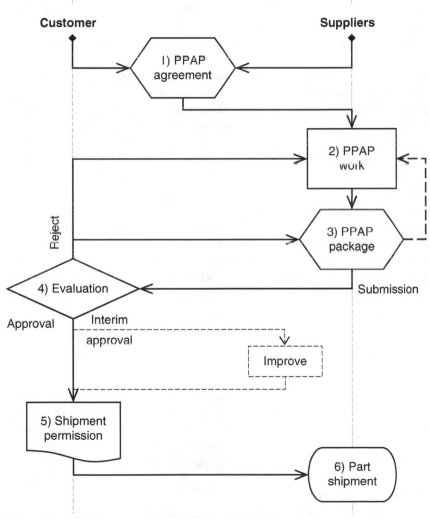

**Figure 6.7** Overall process flow of PPAP implementation.

To start PPAP work, a customer initiates with their preliminary requirements for a particular part (or part family), and communicates with suppliers based on the design requirements of the part. The customer then discusses with its suppliers to aquire a mutual agreement contract as a tentative "first step."

Following PPAP requirements and agreements, suppliers do their work on quality (Step 2) to meet their requirements with support from their customer. During their work, suppliers should document relevant information and prepare a PPAP package (Step 3). The customer reviews the package to validate the supplier's production readiness (Step 4) and then decides to issue a shipment permission for the supplier (Step 5). The major steps in the PPAP process will be discussed further in the sequential subsections.

PPAP agreement submission to the supplier is required for a new part or subassembly. In special cases like significant modifications on the design, specifications, materials, manufacturing processes, tooling, or design records of a part, a PPAP package can also be required. In such cases, the updated PPAP agreement and package are normally only on the elements associated with those modifications. For example, if a design specification is tightened up, then elements related to existing manufacturing, such as a process failure modes and effects analysis (FMEA) and process capability study, are required in the PPAP package. Sometimes, annual reviews on PPAP packages are also required by a customer.

A PPAP is a document package to capture the requirements imposed by a customer, and an approach for suppliers to ensure their readiness and capability for meeting a customers' requirements. Therefore, a PPAP process and its outcomes provide supplier improvement and new opportunities for future business. Industry practices show that suppliers can significantly improve their manufacturing capability by implementing the PPAP process, sometimes going beyond customer requirements.

### 6.2.2 PPAP Elements

**Regular 18 Elements**
Based on the AIAG's guideline, a PPAP has 18 elements, briefly listed in Table 6.1 and further detailed in later subsections. These 18 elements may be grouped into four clusters: 1) engineering design, 2) manufacturing process, 3) quality assurance and records, and 4) summary. These elements are for a manufacturing business; the discussion of PPAP for a service business is in a later subsection.

**Discussion of PPAP Elements**
PPAP requirements, in terms of the elements required, are company and product dependent. Companies often have their own version of PPAP (for examples see Alto-Shaam 2013; GM 2016; FCA 2016; Ford 2017; Cooper n.d.). Although

**Table 6.1** 18 Elements of automotive industry PPAP

| Group | No. | Element | Basic description |
|---|---|---|---|
| 1) Engineering | 1 | Design records | Released designs and specifications; can be provided by either customer or supplier, depending on who handles design. |
| | 2 | Change notice, if any | Description and records of product changes with customer approvals; often from the customer and covered by a PO. |
| | 3 | Customer's approval, if required | Approved deviations and changes from the customer. |
| | 4 | Design FMEA | Completed design FMEA, with recommendations and appropriate signoffs. |
| 2) Manufacturing | 5 | Process flow diagram | A graphical outline of the manufacturing process flow from incoming raw material to shipping, including all steps, with potential quality control points if applicable. |
| | 6 | Process FMEA | Completed process FMEA, with recommendations and appropriate signoffs. |
| | 7 | Control plan | A written description of all inspections for major concerns; often based on process FMEA. |
| | 8 | MSA | Studies performed for measurement devices, including gauge R&R and calibration records. |
| | 9 | Dimensional measurement results | Showing product characteristics, specifications, measurement results, and assessments. |
| | 10 | Material and performance test results | A summary of all results of tests on the part. |
| | 11 | Initial process studies | Evidence of capability through validation and SPC charts for critical characteristics in initial production runs. |
| 3) Quality assurance and records | 12 | Qualified laboratory reports | Laboratory certifications for Element 10. |
| | 13 | Appearance approval report | Applicable for the parts with appearance requirements. |
| | 14 | Sample production parts | Samples from initial production runs. |
| | 15 | Master sample | A good inspected sample, signed off by the customer, often for training purposes. |
| | 16 | Checking aids | Illustrative instructions or tools to inspect or measure parts in production. |
| | 17 | Customer-specific requirements | Specific and additional requirements from the customer. |
| 4) Summary | 18 | Part submission warrant (PSW) | A summary of a whole PPAP package. |

the requirements are demanded by a customer, the final version of a PPAP package is an agreement between the customer and their supplier with additional considerations, such as non-quality factors (e.g. timing and quantity), financial implications, technical capacity, and late engineering changes.

There are natural links among some elements. One example is the connection between design FMEA (#4), process FMEA (#6), and control plan (#7), which is discussed in Chapter 5. Another example may be dimensional measurement results (#9) and master sample (#15). Knowing these relationships may help evaluate quality from different angles, and assist with quality related problem solving.

Besides these standard elements, some companies have additional or different elements. These can be either summarized in Element 17 or themselves additional PPAP elements. A few additional elements, as examples, are listed in Table 6.2. Again, this specific PPAP and its elements are oriented around automotive, high-volume manufacturing. Unique and discipline-specific elements can be required for suppliers in other industries. Each element should have specific requirements, and be formatted based on a customer's guidelines.

In addition to being a quality procedure and requirement, a PPAP itself can be viewed as a toolbox for quality assurance and continuous improvement. For example, the FMEAs and control plans discussed in Chapter 5 are part of a PPAP package as required or recommended elements.

To address specific requirements of service part production, AIAG has a supplement to the production PPAP called service PPAP (AIAG 2014). The service PPAP addresses PPAP expectations for service unique parts, including remanu-

**Table 6.2** Additional elements of company specific requirements

| Group | No. | Element | Description |
| --- | --- | --- | --- |
| Additional | 19 | Customer input | Design criteria and program requirements necessary to start the advanced product quality planning (APQP) process. |
| | 20 | Sourcing selection | Commitment to work with internal and external part suppliers, tooling suppliers, and facility suppliers. |
| | 21 | Subcontractor APQP status | Summarized status of subcontractor (or sub-supplier) as it may affect quantity and quality of supplier's delivery. |
| | 22 | Craftsmanship | Overall, subjective assessment of customer perception by seeing, touching, using, smelling, etc. |

factured parts, service unique chemicals, third party packagers, and service software requirements. Time wise, service parts and their quality assurance should be ready for product launch.

### 6.2.3 PPAP Packages

**Five Levels of PPAP**

The elements of PPAP are grouped into five package levels, roughly according to the levels of significance, complexity, and relevance of elements, for effective management. These PPAP package levels are illustrated in Figure 6.8. Each level has certain requirements from a part supplier. Going from Level 1 through Level 3 of a PPAP package, a supplier is required to submit a few more elements for that PPAP package.

Level 1: Two elements (Elements 18 and 13)
Level 2: Eight elements – product samples and partial supporting data (Level 1 + 6 elements)
Level 3: 16 elements – product samples and complete supporting data (Level 2 + 8 elements)

Level 1 is basic, and applies to simple parts. From Level 1 to Level 3, requirements increase for technical details to support, rather than for quality status. Additional elements are often required for a new, important, and/or complex part. The capability and history of a supplier may also be factors to consider for an element requirement.

For the automotive industry, companies generally start at Level 3. For example, Element 13, appearance approval report (AAR), is not important to many parts. As a result, Level 4 package adjustment is a common practice for most parts. Level 4 comprises the selected elements of Levels 2 and 3, based on the characteristics of parts and the qualification level of part suppliers. A Level 4

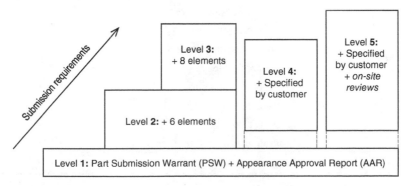

**Figure 6.8** Five levels of PPAP package.

requirement may have fewer elements than Level 3. At Level 5, based on AIAG recommendation, a customer can require additional site visits and reviews as part of a given PPAP package.

Regardless of PPAP package and requirements, a supplier should keep all non-submitted records and documents on site. These documents can be useful for later problem solving and continuous improvement, and are possibly required by a customer. Records should be typically retained for the part's active time plus one year.

### Element Requirements

Similar to the AIAG guideline, many companies have their own guidelines. A typical guideline is shown in Table 6.3. In the table, the elements with a "•" are required for submission at a given level. Other elements are marked as "○" as the supplier shall retain them and make them available upon customer request.

**Table 6.3** Elements and packages of PPAP

| Group | No. | PPAP Element | Level 1 | 2 | 3 | 4/5 |
|---|---|---|---|---|---|---|
| 1) Engineering | 1 | Design records | ○ | • | • | ○/• |
| | 2 | Change notice, if applicable | ○ | • | • | ○/• |
| | 3 | Customer's approval, if required | ○ | ○ | • | ○/• |
| | 4 | Design FMEA | ○ | ○ | • | ○/• |
| 2) Manufacturing | 5 | Process flow diagram | ○ | ○ | • | ○/• |
| | 6 | Process FMEA | ○ | ○ | • | ○/• |
| | 7 | Control plan | ○ | ○ | • | ○/• |
| 3) Quality assurance and records | 8 | MSA | ○ | ○ | • | ○/• |
| | 9 | Dimensional measurement results | ○ | • | • | ○/• |
| | 10 | Material and performance test results | ○ | • | • | ○/• |
| | 11 | Initial process studies | ○ | ○ | • | ○/• |
| | 12 | Qualified laboratory reports | ○ | • | • | ○/• |
| | 13 | AAR, if applicable | • | • | • | • |
| | 14 | Sample production parts | ○ | • | • | ○/• |
| | 15 | Master sample | ○ | ○ | ○ | ○/• |
| | 16 | Checking aids | ○ | ○ | ○ | ○/• |
| | 17 | Customer-specific requirements | ○ | ○ | • | ○/• |
| 4) Summary | 18 | PSW | • | • | • | • |

The requirements for most of the elements at Levels 4/5 are customer specified and case dependent, so they are marked "○/●."

These elements play important roles in quality assurance, but consume resources. Therefore, the selected elements should be justifiable based on several factors, including the efforts and time needed, and the past performance of the supplier. For example, if a supplier has a proven history of reliably in providing quality parts, some elements of a PPAP package can be self certified by the supplier, without a formal review and approval by the customer, or even waived entirely. As long as a customer can be assured on the quality of parts, the PPAP package can be optimally minimized.

## 6.3 PPAP Elements in a Package

### 6.3.1 Essential Element (Level 1)

#### Part Submission Warrant

A PSW (Element 18) is the basic element for any PPAP package regardless of the package level. A PSW is often a two-page document, designed to outline the supplier's responsibility and to support a part in meeting quality and quantity requirements. For packages of Level 2 through Level 5, a PSW serves as a summary of PPAP documents.

AIAG's PPAP manual has a reference form used by the "Detroit Big-three" automakers (AIAG 2006). A customer may directly use the AIAG standard form, or adapt it to a unique PSW form for their suppliers. A PSW form may include the following sections (see Figure 6.9):

1. Part and administrative information: Part name, number, change level, main characteristics, PO, etc.
2. Supplier administrative information: Name, supplier code, address, responsible person, etc.
3. Customer submittal information: Customer name, model year, assembly name, etc.
4. Materials reporting: Material information required by customer.
5. Submission reason: Initial/update, design change, equipment/tooling, source change, etc.
6. Submission details: Related to a PPAP package level.
7. Result summary: Summary of meeting all specifications, to explain if not.
8. Supplier declaration and signature: To sign the warrant with title and contact information.
9. Customer approval: Customer's decision with an authorization signature.

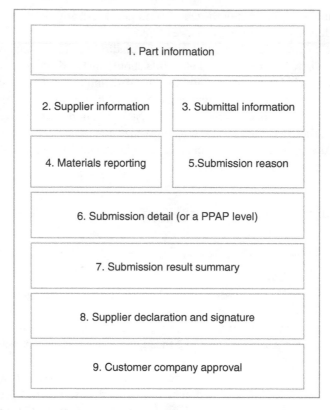

**Figure 6.9** Administrative and information sections of PSW.

A PSW is a formal business document. A supplier should prepare a PSW upon the completion of all PPAP required work, as agreed by both customer and supplier. It is a common practice that a PSW is needed for each part number assigned by the customer. For a larger quantity, a part may be produced from multiple pieces of machinery and processes. In such a case, a supplier should include the data from each machine and process in a PSW report.

**Considerations for Level 1**
A Level 1 PPAP package consists of at most three documents. The basic one is PSW discussed above. The other two may be needed if they are applicable. If a part is visible to customers, Element 13, AAR, is normally required. For some product subassemblies, substances called "bulk materials" – namely adhesive, sealants, chemicals, coatings, fabrics, lubricants, etc. – are used. For the suppliers who provide such materials, their information, quality status, and expiration time requirements, etc. are required. AIAG considers a bulk material quality to be a Level 1 submission requirement conformance of PPAP (AIAG 2006).

Element 13 – AAR is part of a Level 1 package submission, but only as applicable. An AAR is concerned with aesthetic requirements, such as color, grain, or finish. An AAR is applicable only for some parts, and is unnecessary if a part is not visible on the product by customers. For automotive vehicles, an AAR is required for all interior, exterior, luggage compartment, and selected under-hood components, as they are visible to customers. Sometimes, a visual "match-to-master" inspection is conducted for an appearance evaluation.

As mentioned, the quality related information and some specific requirements of "bulk materials" may be needed. After being assigned a part number, a bulk material supplier should go through a PPAP process. AIAG also has a recommended checklist for the quality of bulk materials.

## 6.3.2 Level 2 Elements

At Level 2, in addition to Elements 18 and 13, six supporting elements are recommended.

### Engineering Design (Elements 1 and 2)

- Element 1 – design records. This document includes all part design files, CAD/CAM data, characteristics, specifications, bill of materials, and related interface (to neighboring parts) requirements. These records shall be released and authorized for production and/or service use. The responsibility of design records depends on who designs the part, either the customer or supplier.
- Element 2 – change notice (if any). A change notice is the authorized document for a new change to be incorporated in a product, as the results of design changes. Such changes occur after the initial PPAP agreement. Therefore, a change notice needs to be covered by a new PO to the supplier from the customer if applicable. Change management will be discussed in Chapter 7. It is common to have multiple design changes during the supplier's work on manufacturing planning. It is crucial that the latest change order is implemented.

### Quality Assurance and Records (Elements 9, 10, 12, and 14)

- Element 9 – dimensional measurement results. The results are artifacts of dimensional verifications and of a part meeting dimensional requirements. When possible, a supplier should measure a five-part sample for reliable data. If a part is produced in multiple manufacturing processes, measurement results are needed for each process. This element report should include all relevant information (like in Figure 6.10). Measured part samples are normally saved for future reference and problem-solving review.
- Element 10 – material and performance test results. Material and/or performance tests apply to some suppliers, particularly to a supplier who provides an

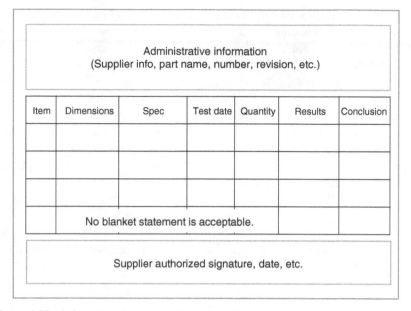

**Figure 6.10**  Information in product dimensional test report.

entire functional module, or raw materials. This element is required for bulk materials as well. Material test requirements, such as on mechanical, electrical, chemical, or metallurgical characteristics, are specified by the customer. Performance tests are product dependent, such as torque, flow, pressure, endurance, and packaging. Test samples are normally saved as an evidence and reference. The material test report is prepared in a similar format as others (see Figure 6.11). In the automotive industry, reports on a form of DVP&R (design verification plan and report) mentioned in Chapter 4 are often used.

- Element 12 – qualified laboratory reports. A laboratory report certifies the qualification of a test laboratory, maybe an external or commercial one, for the tests conducted. A laboratory should comply with the latest ISO/IEC 17025 requirements (ISO 2017). An accreditation of ISO/IEC 17025 may be required, depending on the customer.
- Element 14 – sample production parts. Samples should be from a production trial run. In a report, the sample size, serial numbers, or other methods used to identify individual parts should be included, along with physical samples. The quantity of sample parts, often at least three pieces, is specified in a PPAP package by the customer.

It is important that all records and sample parts represent the actual, "typical" conditions for a supplier. Using only the best parts and/or data makes

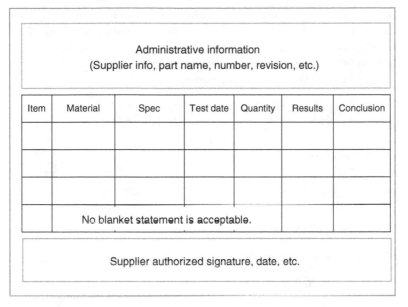

**Figure 6.11** Information in material and performance test report.

the PPAP files look good, but hides the actual issues, which will really hurt the supplier in the future.

### 6.3.3 Level 3 Elements

Moving up to package Level 3, additional eight elements from Level 2 are suggested.

#### Engineering Design (Elements 3 and 4)

- Element 3 – customer's approval (if required). This is documentation, consent, interpretation, and/or a waiver provided by the customer as an official confirmation to adopt a particular change, which may be on a product, requirement, material, etc.
- Element 4 – design FMEA. It is needed by the customer if the supplier is responsible for part design. The process and technical contents of FMEA are discussed in Chapter 5.

#### Manufacturing Process (Elements 5, 6, and 7)

- Element 5 – process flow diagram. This diagram is a graphical representation of the entire manufacturing process from material receiving through part shipping, including material handling and transportation. Figure 6.12 shows a partial process flow as an example. The diagram should include the corresponding quality process, such as inspections, tests, and scrap/reworks for the part. Product characteristics, process characteristics, and other PPAP elements can be marked on a process diagram as needed.

- Element 6 – process FMEA. As a proactive approach, it is strongly recommended that a supplier do a process FMEA. This element should be aligned with Element 5 – process flow diagram. The details of a process FMEA are discussed in Chapter 5.
- Element 7 – control plan. A control plan is required for key functions, features, and characteristics. The details of a control plan are discussed in Chapter 5. This element should be aligned with the Element 5 – process flow diagram and Element 6 – process FMEA.

### Quality Assurance and Records (Elements 8 and 11)

- Element 8 – MSA. This document shows that the capability of instruments used for quality measurements has been studied, and meets requirements. The details of MSA process and criteria are discussed in Chapter 7.
- Element 11 – initial process studies. This record is about the quality capability of a manufacturing process. The details of a process capability study and criteria are discussed in Chapter 7.

In a Level 3 package, 16 out of the 18 elements are included. There are two remaining elements: 15 and 16. Element 15 – master sample, which is like

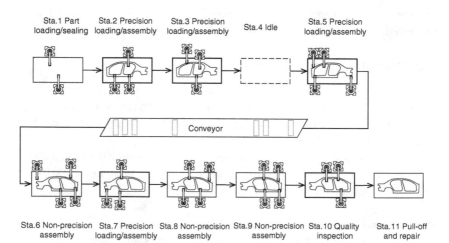

**Figure 6.12** Example of manufacturing process flow diagram.

Element 14 – sample production parts, includes a masterpiece of the best part selected. A masterpiece should be kept as a record for a period, and may be used for training purposes. Element 16 – checking aids may be required when a special quality inspection procedure or method is used for a part during production. In some PPAP guidelines, Elements 15 and 16 are included in a Level 3 package.

### 6.3.4 Unique Requirements (Levels 4 and 5)

Even with similar parts, different companies can have different requirements for a PPAP package. Element 17 is about customer-specific requirements. These are flexible for a PPAP package, because the item selected for this element is based on the part and customer concerns. Examples include:

- APQP project timeline
- Action log
- Launch plan
- Production feasibility agreement
- Production readiness review
- Gauge development plan
- Dimensional correlation matrix
- Packaging specification data
- Barcode label packaging
- PPAP interim recovery plan

Additionally, a Level 4 package can include any combination of the regular 18 elements. Thus, it is the most common, practical choice for PPAP applications. For example, Volvo Cars requires Elements 5 (process flow), 6 (process FMEA), 7 (control plan), 9 (dimensional measurement results), 10 (test results), 13 (appearance report), and 18 (PSW) for most physical parts in its PPAP packages (Volvo 2019).

At Level 5, a customer conducts a review on the supplier site as part of a PPAP package. A supplier provides the required PPAP elements, and prepares for an on-site review at its production location. An on-site review is a common practice of SQM, particularly for important parts and subassemblies. It often accompanies a real-time production run of the parts; therefore the supplier can demonstrate the readiness and quality of a part to support the customer's production. Supplier representatives should respond quickly to any major issue happening during an actual run. An on-site review also assists with communication between a customer and supplier for PPAP approvals.

## 6.4  Supplier Quality Assurance

### 6.4.1  PPAP Preparation and Approval

**PPAP Preparation**

A supplier takes effort and time for PPAP preparation. Supplier management needs a plan for its PPAP preparation, not only for timing but also for resources. Depending on the requirements of PPAP and complexity of parts, one PPAP preparation can take several weeks to complete (Rydström and Viström 2020). It is a good idea to create and use a checklist based on PPAP requirements to prepare PPAP documents. Table 6.4 shows a partial example. If some data and files are missing or not satisfactory to support PPAP requirements, a follow-up and additional work are needed to complete a PPAP package.

If the PPAP elements are an integral part of the supplier's QMS, the preparation work and respective documents may have already been completed internally. In this case, a supplier project manager mainly compiles existing files for PPAP package preparation, which would not need too much extra work.

PPAP data must be collected from a production run. All data shall reflect the actual production conditions, meaning a supplier shall use:

- Actual production processes, equipment, and tooling
- Actual production employees, while running at the production rate
- Running this actual system and producing 300 consecutive parts (or the quantity determined by the customer)

Supplier management often plans and executes separate runs, in addition to a PPAP package. If all results conform, then the supplier proves their capabilities

**Table 6.4**  Example of PPAP preparation checklist

| | | Administrative Section (Customer, project, manager, time, etc.) | | |
|---|---|---|---|---|
| **No.** | **PPAP element** | **Satisfactory** | **If no, follow-up** | **Remarks** |
| | | **Yes/no** | **Who**     **When** | |
| | ... ... | | | |
| 7 | Control plans | | | |
| | 7a. Control plan for Part D | | | |
| | 7b. Control plan for Process 3 | | | |
| | 7c. Control plan for Material M | | | |
| | ... ... | | | |

and can finalize all required documents. However, if there are any nonconforming observations or issues, the supplier needs to develop a recovery plan and direct actions. A recovery plan and associated actions follow the same process for root cause analysis, problem solving, variation reduction, and continuous improvement. For example, a supplier does a trouble-shooting analysis to find the root causes. Then, they need to take corrective actions, and re-run the process until success. A supplier should communicate with its customer about main quality issues and improvement projects. A PPAP package should be fully completed only after all issues are resolved. Readers can refer to Chapter 8 for problem-solving processes and quality management tools for such improvement actions.

### PPAP Approval

After receiving a supplier's PPAP package, the customer reviews the PPAP paperwork and examines the samples produced, based on the requirements described in the PPAP agreement. This evaluation consists of a production readiness audit, review of any concerns, cost analysis, etc. There are three possible outcomes of an evaluation: approval, interim (or conditional) approval, and rejection (see Figure 6.13).

1. An <u>approval</u>, as an official document, indicates that the part and documentation fully meet all specifications and requirements. The supplier can ship production quantities of the part to the customer.
2. An <u>interim approval</u> means that the PPAP package is not fully satisfactory and/or sample parts do not fully meet the requirements. However, the customer determines that the quality of a part is marginally acceptable. A shipment permit for the part is granted for a limited time (with an expiration date) and on a specified quantity for the part. In the meantime, the supplier must have an improvement plan and quick reactions to obtain full approval.

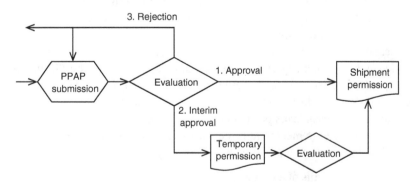

**Figure 6.13** Approval process flow of a PPAP package.

3. A <u>rejection</u> states the PPAP package and/or sample parts do not meet customer's requirements, based on the production lot from which it was taken and/or accompanying documentation. The supplier cannot ship parts and must improve their quality to meet the requirements within a defined deadline. A rejection is a serious situation, because it can affect the entire new product development of a given customer.

The PPAP evaluation and approval processes can be challenging for a customer, given the many requirements and PPAP elements for a part, and PPAP packages to be reviewed during a short time window. A new product may have hundreds or even thousands of parts. It is a challenge for both suppliers and customers to have quality assurance completion and approvals for all parts, including service parts, on time. For example, an automaker had one launch delayed by a few months because of product quality concerns, a lack of adequate supplier preparations, etc. (Berhausen and Hannon 2018).

In addition, to make an on-time launch or delivery, quality issues can have significant impacts on cost. An interim approval means additional work for a supplier in a short period. As for an official approval, any missing items or quality issues of a PPAP package, after an approval, can be the responsibility of the customer. If needed, amendments to an approved PPAP file are required for the items in question. Any additional work after an approval, such as quality improvement or specification modification, often has financial implications for the customer.

### 6.4.2 Customer and Supplier Teamwork

**Customer Leadership**

When a supplier involves themselves with the business operations of a customer, the supplier becomes partnered with that customer. From a quality perspective, suppliers are an extension of customer enterprise, rather than a separate entity. A customer wants high quality, reliable delivery, and low price, while a supplier expects a stable/growing business from their customer, and healthy profits. A customer and its supplier share some common interests, i.e. business success and ultimate customer satisfaction.

Successful partnership is effective to achieving the goals of a customer and its supplier, which is conceptually discussed in Chapter 1. Like the Philips' supplier quality mission states, "Together we are inspired to enable best in class value chain partnerships to the delight of our customers" (Philips, n.d.).

A customer always leads partnership development, PPAP development, and supports a supplier to meet requirements. To develop a partnership, a customer can follow these seven steps (Russ 2013):

1. Determine benefits of a partnership
2. Understand and consider legal implications

3. Identify suppliers for a partnership
4. Define expectations and potential risks
5. Communicate and negotiate with suppliers
6. Implement the mutually agreed arrangements
7. Establish constant monitoring and communication

The first step is critical, to motivate a customer to think about and initiate partnership development. The main benefits of a partnership are that a company can attain better quality assurance, with fewer quality issues of suppliers, and production will be less interruptive. Benefits also include better pricing for customers, more investment in innovation and technology, and better supplier support. A recent study, called the Annual North American Automotive OEM–Supplier Working Relations Index Study, showed that Toyota was indexed at the highest (345) and Honda (310) for the automotive OEM–supplier work relationship, while FCA scored 198 and Nissan 190 (Irwin 2020).

A customer normally has a dedicated SQM team, who helps suppliers improve their performance with orders, requirements, POs, and deliverables. The customer SQM team plays a lead role on the supplier selection during the sourcing process, delivering successful supplier launch of new parts, monitoring supplier quality, and manufacturing systems sustainability, etc. Such an SQM team is titled as supplier quality engineering, supplier quality assurance, or supplier development engineering in manufacturing industries among others. For example, the SQM at the Ford Motor Company is called Supplier Technical Assistant (STA).

The partnership practice has a large variation across industry: some companies involve suppliers' quality management on site, while others just perform verifications and approvals. The practice variation also depends on different factors, including corporate culture, policy, importance of parts, and the proprietary information of suppliers. Without patent protection and/or licensing agreements, a supplier should not be asked to disclose its proprietary information.

**Roles in Partnership**

If following the PPAP principle, a customer and its suppliers play different roles for individual PPAP elements. Table 6.5 shows a notional example. For a specific case or part, the actual roles of supplier and customer can be different.

A supplier ensures good parts by fully inspecting those parts before shipping to its customer, based on the standing PPAP and associated agreements. This allows the customer to use those parts with no further checks. If the received parts have quality issues, the consequences at a customer site, e.g. production downtime and extra repair work, can result in a financial burden to the supplier, as defined in the contract. Such a quality assurance guarantee makes

**Table 6.5** Roles of customer and supplier in PPAP development

| Group | No. | PPAP element | Supplier to | Customer to |
|---|---|---|---|---|
| | 1 | Design records | Depending on design responsibility | |
| 1) Engineering | 2 | Change notice, if any | Receive | Provide |
| | 3 | Customer's approval, if required | Request | Provide |
| | 4 | Design FMEA | Lead | Participate |
| | 5 | Process flow diagram | Provide | Review |
| 2) Manufacturing | 6 | Process FMEA | Lead | Participate |
| | 7 | Control plan | Generate | Review |
| | 8 | MSA | Generate | Review |
| | 9 | Dimensional measurement results | Generate | Review |
| | 10 | Material and performance test results | Generate | Review |
| 3) Quality assurance and records | 11 | Initial process studies | Generate | Review |
| | 12 | Qualified laboratory reports | Provide | Receive |
| | 13 | AAR, if applicable | Generate | Review |
| | 14 | Sample production parts | Select | Receive |
| | 15 | Master sample | Select | Review |
| | 16 | Checking aids | Generate | Receive |
| | 17 | Customer-specific requirements | Receive | Provide |
| 4) Summary | 18 | PSW | Generate | Review |

supplier business risky. Thus, the partnership motivates a supplier to work diligently toward quality assurance, and closely with its customer, to have consistently good quality, share efforts, risks, and rewards.

To support a supplier partnership, the roles and responsibilities of all associate departments inside a customer organization should be defined. In general, the management team of supplier quality coordinates the processes, e.g. PPAP, and the responsible department plays a lead role, while others support. The support roles may be specified as input, participation, or concurrence, etc. Table 6.6 shows an example.

As discussed at the beginning of this chapter, SQM also includes the monitoring and control of supplier quality during regular operations. In other words, having a successful supplier quality assurance, like PPAP, is a good starting

**Table 6.6**  Role matrix of customer company departments for PPAP

| No. | Element | Department | | | | | |
| --- | --- | --- | --- | --- | --- | --- | --- |
| | | Project | Product | Manuf. | Production | Purchasing | Supplier |
| ... | | | | | | | |
| 8 | MSA | | Support | Lead | Support | | |
| ... | | | | | | | |
| 19 | Customer input | Lead | Input | Input | Participate | Participate | |

point for SQM. Continuous monitoring, control, and improvement efforts from the partnership are needed after product launch.

**Toyota Supplier Quality Assurance**

There are different processes in supplier quality assurance relative to PPAP, but these processes are based on identical or similar principles and requirements. Many automotive and non-automotive manufacturing companies have similar guidelines, called Supplier Quality Assurance Manual (SQAM) to specify the process and requirements for their suppliers. A typical content of an SQAM file is shown in Figure 6.14.

For production part approval by Toyota, the requests include (Iyer 2015):

- Project plan with major milestones
- Supplier part master plan
- Process capability studies

1. General Requirements
   ...
2. Sourcing Selection
   ...
3. Business Partnership Process
   3.1 Plan
   3.2 Responsibilities
   3.3 Contract reviews
   3.4 Sub-supplier management
   3.5 Change management
4. Supplier Requirements
   4.1 General guidelines
   4.2 Sample requirements
   4.3 PPAP elements applicability

5. Serial Production Requirements
   5.1 Early production exit criteria
   5.2 Operating procedures
   5.3 Tests, reviews, and verifications
   5.4 Packaging and labelling
   5.5 Deviation request and approval
   5.6 Record retention
   5.7 Warranty
6. Supplier Performance Measurement
   6.1 Managing processes
   6.2 Corrective action reporting
   6.3 Supplier assessment
   6.4 Supplier audit and ratings
   6.5 Continuous improvement

**Figure 6.14**  Main contents of SQAM.

- Process FMEA or DRBFM
- Part evaluation plan and inspections
- Manufacturing quality charts
- Sample part submission
- Part final approval

Toyota's SQAM emphasizes not only the technical details of part quality but also the corresponding project management of suppliers. Project timing for each part supplier should be based on Toyota parts' master schedule. The general timing of parts has four preparation phases, with defined supplier responsibilities (Figure 6.15):

Phase I: <u>Planning</u>, to have clear expectation and understanding of quality requirements. Tasks include establishing quality requirements, quality control standards and criteria, conducting risk assessment, FMEA, concurrent engineering activities, etc.

Phase II: <u>Initial evaluation</u>, to achieve the quality requirements on production tooling. Tasks include production tooling, process capabilities, inspection plan and criteria, and related analyses, samples, and documents.

Phase III: <u>Final verification</u>, to ensure consistent quality for volume production. Tasks include final submission of all required assurance documents, all trials completed, corresponding verifications, and approvals.

Phase IV: <u>Mass production launch</u>, to ensure quality through launch and production volume ramp-up. Tasks include inspections, monitoring, adjustments, final approval, and lessons learned documentation.

For part quality inspections, Toyota has the Toyota Inspection Standard (TIS) sign-off, which includes all the inspection requirements, including functional, dimensional, and visual, for a given part. Suppliers may be requested to provide similar documents, such as a control plan, inspection agreement, test results, and dimensional measurement results. In addition, Toyota normally conducts two prototype trials for parts, with an increased volume the second time, and has a sign-off process and document for each run.

With these four phases and multiple trials, a supplier would review step-by-step and submit the required documents multiple times, while a PPAP package

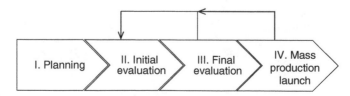

**Figure 6.15**   Overall process flow of SQAM.

normally requires one submission. This planned, multiple-phase customer–supplier teamwork may be a better way to achieve and sustain the good quality of suppliers.

### 6.4.3 Supplier Quality to Service

**Service Supplier Quality**

Service providers themselves need suppliers for parts, services, and technical support. For example, a healthcare organization often outsources some non-core operations, e.g. IT support, main equipment maintenance, and janitorial duty, to external contractors (or suppliers). However, there are some differences between service providers and product producers for SQM.

The major differences between product development and service development include their outcomes and their characteristics. Chapter 1 shows that service quality has five dimensions (refer to Table 1.3): reliability, responsiveness, assurance, empathy, and tangibility. Unlike a product, most service functions and their quality characteristics are intangible, not only for customers but also for development and management teams. The development process and characteristics for a service differ from those of a product. In addition, the service sector itself, and suppliers for services, are even more diverse than those of the automotive industry.

Conventional PPAP is for the quality assurance of production parts in manufacturing industries, as specific elements in a PPAP package are specifically designed for physical parts. The principle and process of PPAP can apply to service industries, albeit not in the strictest sense; a service, as a "soft" product and its diversity, is still packaged into a systematic guide like PPAP.

To ensure the performance and quality of core business operations, a service provider must carefully select capable and reliable suppliers. The PPAP principle and process may help a service organization manage supplier quality. However, the detail items of PPAP may not directly apply to service sectors, and the measurements of service quality suppliers can be different. Adopting the PPAP approach in the service industry is an interesting study subject for service quality professionals.

To develop the quality assurance process and package for a supplier, one may use the PPAP as a reference. At the same time, bear in mind the differences between a part and service, one may look at the PPAP elements to check whether they may apply to service quality assurance.

**Discussion of Service Supplier Quality**

Table 6.7 lists the possible applicable warrant elements for a generic service, in a PPAP format. Some elements are applicable to service, some are not, while others fall in between. Even though a PPAP element is applicable, its name will

**Table 6.7** Possible PPAP application to service industry

| Group | No. | PPAP element (product) | Applicability | Service quality assurance |
|---|---|---|---|---|
| Design 1) Engineering | 1 | Design Records | Yes | Same functionality |
| | 2 | Change notice, if any | Yes | Same functionality |
| | 3 | Customer's approval, if required | Yes | Same functionality |
| | 4 | Design FMEA | Yes | Service FMEA, similar functionality |
| Procedure 2) Manufacturing | 5 | Process flow diagram | Yes | Similar functionality on service |
| | 6 | Process FMEA | Maybe | For service procedures |
| | 7 | Control plan | Yes | Same functionality |
| 3) Quality assurance and records | 8 | MSA | Maybe | Depending on the type of service |
| | 9 | Dimensional measurement results | Yes | Similar, e.g. service measurements |
| | 10 | Material and performance test results | No | |
| | 11 | Initial process studies | Yes | Similar, e.g. service tryout reports |
| | 12 | Qualified laboratory reports | Maybe | Depending on the type of service |
| | 13 | AAR, if applicable | Maybe | Samples, depending on the type of service |
| | 14 | Sample production parts | Yes | Similar, e.g. initial case studies |
| | 15 | Master sample(s) | No | |
| | 16 | Checking aids | No | |
| | 17 | Customer-specific requirements | Yes | Same functionality |
| 4) Summary | 18 | PSW | Yes | Similar functionality, service warrant |

be different for service, as opposed to product applications. Note Table 6.7 is just an example, meant to inspire creative thought and discussion. Similar to the automotive PPAP, the selection of the elements in a service warrant is determined by a customer service provider.

For service, the first group can be renamed design from engineering. Group 2 is retitled procedure, instead of manufacturing, as these elements in the group are likely used for service processes and procedures. The elements in the third group may be less straightforward for services; their applicability varies depending on the characteristics and features of a service. Group 4 is a single element of a summary, which can be called a service warrant. One might agree that the PPAP is a useful reference to develop supplier quality assurance in service sectors. There are additional service-specific items to be addressed, beyond those considered so far.

Overall, following the same process, the quality review and approval process for a service supplier can be developed like a product supplier. The fundamental element therein is a service quality warrant. Additional elements may be created for a specific service. The "production part," the first two Ps in the PPAP, could be replaced with "service." The manufacturing-orientated quality assurance guidelines, e.g. Toyota's SQAM, could be another reference for the guideline development of service supplier quality.

## Summary

### Introduction to Supplier Quality

1 The significance of suppliers to a customer can be categorized into three levels of significance: essential, important, and supportive.
2 SQM includes four stages: planning, sourcing selection, quality assurance, and monitoring and control.
3 In the process of sourcing selection, a customer considers both technical and financial aspects.
4 AHP is a structured approach to analyze supplier candidates based on major factors, in quality related and other aspects, with various weights.
5 Certification, monitoring, and change management are the main follow-up tasks of SQM for the selected suppliers.

### PPAP Standardized Guideline

6 PPAP is a quality-assurance approach, which was initially developed for the suppliers in the automotive industry and is applicable in other manufacturing industries.
7 PPAP is a joint process and effort, between a customer and its suppliers.
8 A complete PPAP has 18 elements in four clusters: 1) engineering design, 2) manufacturing process, 3) quality assurance and records, and 4) summary.

**9** Additional and special elements of a customer may be added to a regular PPAP.

**10** The elements of PPAP are arranged into five package levels, depending on customer's requirements and complexity of parts.

### PPAP Elements in Package

**11** A PSW is the basic element for any PPAP package.

**12** A PPAP Level 2 package includes six additional elements to Level 1.

**13** A PPAP Level 3 package includes eight additional elements to Level 2.

**14** PPAP Level 4 is a flexible and unique package, including selected elements from the 18 ones and/or additional elements.

**15** PPAP Level 5 is similar to Level 4, except that it also requires on-site reviews.

### Supplier Quality Assurance

**16** PPAP preparation and submission are supplier responsibility, with the customer's assistance.

**17** PPAP approval is based on the PPAP agreement between a customer and supplier.

**18** An approval decision has three types: approval, interim approval, and rejection, as determined by the customer.

**19** Partnership and teamwork between a customer and supplier with defined roles, responsibilities, and collaboration is effective to PPAP success and supplier quality assurance.

**20** SQM in service sectors differs from manufacturing, while PPAP principle and process can be useful references.

## Exercises

### Review Questions

**1** Talk about three major components of SQM with an example.

**2** Explain the roles of suppliers to the ultimate customers through a customer.

**3** Discuss the two types of internal review for a source/supplier selection.

**4** Discuss factors to consider for a source/supplier selection.

**5** Discuss an approach application of supplier quality monitoring.

**6** Explain the purposes of PPAP.

**7** Discuss the overall process of PPAP.

**8** Review one cluster of PPAP elements with an example.

9 Explain the five levels of PPAP packages.

10 Propose a new PPAP element in addition to the regular 18, for a part's quality.

11 For a part with a critical characteristic, which level of PPAP package is likely to be used? Why?

12 Discuss some considerations during PPAP preparation and implementation.

13 Discuss the necessity of an on-site review for a product part or service function.

14 Describe the three types of PPAP approvals.

15 List and explain the considerations to grant an interim approval to a PPAP package.

16 Discuss the partnership between customer and suppliers, when applying the PPAP approach.

17 Review a difference between PPAP and SQAM.

18 Select a dimension of service quality and propose to relate it to a PPAP element.

19 Select a PPAP element and discuss its applicability to service.

20 Propose the key information of PSW if it applies to a service supplier.

## Mini-project Topics

1 There is an argument that selecting suppliers based on the lowest upfront cost is likely problematic. Do you agree or disagree with this statement? Why?

2 Search for and review a successful application of PPAP.

3 Find an application of PPAP, and review which elements are required in the PPAP package.

4 Find an application of PPAP, in a non-automotive industry, and analyze its uniqueness.

5 Propose and justify a new element, in addition to the 18 PPAP elements, for an application.

6 Suppose you are coming into ownership of a family catering business, or a new startup service (of any other type). To ensure service quality, you are considering the implementation of the PPAP approach. Please develop a simple/overall plan of PPAP implementation, e.g. what level and which elements, with brief introduction and justification.

7 If you revise the PPAP guideline and extend it to general manufacturing from automotive manufacturing, do you propose to remove any of the 18 elements? Would you suggest new ones?

8 Search for and study the benefits of the partnership between a customer and its suppliers, with examples.

9 Study the supplier management guideline of a company (e.g. SQAM), and review its PPAP-like elements.

10 Review the applicability of the PPAP approach for a service industry.

# References

AIAG. (2006). *Service Production Part Approval Process (Service PPAP)*. Southfield, MI: Automotive Industry Action Group.

AIAG. (2014). *Production Part Approval Process PPAP*, 4th edition. Southfield, MI: Automotive Industry Action Group.

Alto-Shaam. (2013). Introduction to Production Part Approval Process (PPAP), http://supplierportal.alto-shaam.com/Portals/3/Supplier_PPAP_Manual_71813. pdf, accessed in June 2018.

Berhausen, N. and Hannon, E. (2018). Managing change and release, March 20, 2018, McKinsey & Company, https://www.mckinsey.com/business-functions/operations/our-insights/managing-change-and-release#, accessed in September 2020.

BioPham. (2020). FDA finds CGMP violations at Canadian facility, *BioPharm International Editors*, https://www.biopharminternational.com/view/fda-finds-cgmp-violations-at-canadian-facility, accessed in October 2020.

Cooper. (n.d.). PPAP submission requirements quick reference, http://www.cooperindustries.com/content/dam/public/Corporate/Company/Sourcing/Cooper_Industries_PPAP_Quick_Reference.pdf, accessed in June 2018.

FCA. (2016). *Customer-specific requirements for PPAP, 4th edition and service PPAP*, 1st edition, Fiat Chrysler Automobiles, Publication Date: October 14, 2016.

Ford. (2017). *Customer-specific requirements for IATF-16949:2016*, Ford Motor Company, Effective May 1, 2017.

GM. (2016). *GM customer specifics – ISO/TS 16949, Including GM Specific Instructions for PPAP 4th edition*, General Motors Global Supplier Quality, Effective Date: October 2016.

Hosseini, S. and Khaled, A.A. (2019). A hybrid ensemble and AHP approach for resilient supplier selection, *Journal of Intelligent Manufacturing*, Vol. 30, pp. 207–228. 10.1007/s10845-016-1241-y.

IATF. (2016). *IATF 16949 Quality management system requirements for automotive production and relevant service parts organizations*, International Automotive Task Force.

Ireland, W.C. (2017). Selection of an alternative production part approval process to improve weapon systems production readiness, Master Of Science In Engineering Systems, Naval Postgraduate School, Monterey, CA. https://apps.dtic.mil/sti/citations/AD1046854, accessed in October 2020.

Irwin, J. (2020). Toyota sweeps annual OEM-supplier relations study, *Wards Auto*, June 22, 2020. https://www.wardsauto.com/industry/toyota-sweeps-annual-oem-supplier-relations-study, accessed in August 2020.

ISO. (2015 a). *ISO 9001: 2015Quality Management Systems – Requirements*. Geneva: International Organization for Standardization.

ISO. (2015 b). *ISO 14001:2015 Environmental Management Systems – Requirements with Guidance for Use.* Geneva: International Organization for Standardization.

ISO. (2017). *ISO/IEC 17025 General Requirements for the Competence of Testing and Calibration Laboratories.* Geneva: International Organization for Standardization.

Iyer, R.L. (2015). Toyota Part Approval (PA) Process, https://www.slideshare.net/RahulIyerMSEMechEngE/toyota-part-approval-pa-process, accessed in August 2020.

Manivel, P. and Ranganathan, R. (2019). An efficient supplier selection model for hospital pharmacy through Fuzzy AHP and Fuzzy TOPSIS, *International Journal of Services and Operations Management*, Vol. 33, Iss. 4, 10.1504/IJSOM.2019.101588.

NHTSA. (2001). Firestone tire recall, https://one.nhtsa.gov/Vehicle-Safety/Tires/Firestone-Tire-Recall, Updated 10/ 04/2001, accessed in August 2020.

Philips. (n.d.). Supplier quality manual, Philips, https://www.philips.com/c-dam/corporate/about-philips/sustainability/supplier-quality-manual-v1.pdf, accessed in October 2020.

Queensland. (2020). Identifying supply chain risks, Last updated: June 10, 2020, The State of Queensland, https://www.business.qld.gov.au/running-business/protecting-business/risk-management/supply-chains/identifying, accessed in October 2020.

Russ, W., ed. (2013). *The certified manager of quality/organizational excellence handbook*, 4th edition., Milwaukee, Wisconsin: Quality Press

Rydström, A. and Viström, J. (2020). Production part approval process evaluation – A case study at a large OEM, Report No. E2019:127, Chalmers University Of Technology, Gothenburg, Sweden.

SAE. (2016). *Aerospace Series–Requirements for Advanced Product Quality Planning and Production Part Approval Process AS9145.* Warrendale, PA: SAE International

Volvo. (2019). Supplier quality assurance manual, 5th edition, 03-2019. https://www.volvogroup.com/content/dam/volvo/volvo-group/markets/global/en-en/suppliers/our-supplier-requirements/SQAM-2019.pdf, accessed in May 2020.

Yan-Kai Fu, Y. (2019). An integrated approach to catering supplier selection using AHP-ARAS-MCGP methodology, *Journal of Air Transport Management*, Vol. 75, pp. 164–169, 10.1016/j.jairtraman.2019.01.011.

Zeng, J., Phan, C.A., and Matsui, Y. (2013). Supply chain quality management practices and performance: An empirical study. *Operations Management Research*, Vol. 6, Iss. 1–2, pp. 19–31. 10.1007/s12063-012-0074-x.

# 7

# Special Analyses and Processes

Quality planning and plan execution involve many analyses and processes. Four important ones discussed in this chapter are: 1) measurement system analysis (MSA), 2) process capability study, 3) development change management, and 4) quality auditing. To be proactive in quality management, these analyses and processes should be planned for, and conducted during, the development phases of a new product or service.

## 7.1 Measurement System Analysis

### 7.1.1 Measurement System

#### Understanding of Measurement

Measurement, as an essential part of quality management, quantifies quality status and change over time for a product, process, or service. In most cases, quality is not manageable without measurement. Performing measurements and analyzing their results are a foundation for understanding the quality of a product, process, or service and can lead to a better understanding of current status and improvement opportunities. Therefore, instrumental measurement of inputs, settings, processes, efforts, and outputs, is an integral part and supporting foundation of a business operation, shown in Figure 7.1.

The American Society for Quality (ASQ) defines "measurement" to be "the act or process of determining a value. An approximation or estimate of the value of the specific quantity subject to measurement, which is complete only when accompanied by a quantitative statement of its uncertainty" (ASQ n.d.). In the healthcare industry, "Quality measurements typically focus on structures or processes of care that have a demonstrated relationship to positive health outcomes and are under the control of the health care system" (AHRQ n.d.).

*Quality Planning and Assurance: Principles, Approaches, and Methods for Product and Service Development*, First Edition. Herman Tang.
© 2022 John Wiley & Sons, Inc. Published 2022 by John Wiley & Sons, Inc.

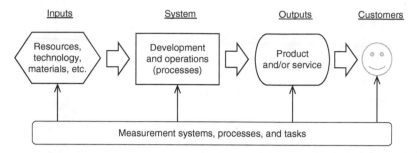

**Figure 7.1**   Roles of measurement in development and operation.

Quality measurements can be either quantitative or qualitative. That is, the numeric expressions of an object on its continuous variables based on a fixed reference (unit) or the non-numeric expressions of an object on its attribute variables. For example, the quality of a product can be measured by a rate of defects, e.g. the number of quality problems per 100 vehicles on average for a new car model. The quality of a healthcare institute can be measured both qualitatively – e.g. ranking in a geographic region or specialty – and quantitatively – e.g. patient retention rate.

A measurement system is a summary of principles, procedures, methods, instruments, and human appraisers involved in making a given measurement. In the development process of a product or service, quality measurement systems should be ready earlier for use, e.g. during prototype and test phases.

Note, all measurements come with certain errors. A measurement error is an uncertainty, the difference between a measured value and the true value. Measurement errors are unavoidable, but controllable in most cases. The capability of a measurement system must be known, as its inaccurate measurements can even be misleading.

**Source of Measurement Error**

To understand measurement errors, one must test, calibrate, and analyze a measurement device, instrument, or equipment before using it for a quality assessment. The processes to evaluate measurement errors and qualify a measurement system are called measurement system analysis (MSA). As a quantitative verification, an MSA shows the uncertainty of a measurement system and its relation to the objects being measured. Other types of verification process, e.g. a quality audit, may be used for a measurement system, but maybe in a qualitative sense.

As an assurance to quality measurement, MSA should be planned and completed in a quality planning stage. For example, MSA is an important part of advanced product quality planning (APQP) discussed in Chapter 1. Furthermore, MSA can be a routine requirement and effort during operations to ensure the correctness of measurements and monitoring for quality work and customer satisfaction.

There are two basic factors associated with measurement errors: one is the measurement instrument itself, and the other is the people who use the instrument. The former includes accuracy and repeatability, while the latter is about reproducibility.

- Accuracy: It describes the difference between the measurement average and the true value, which is often presented in a range, e.g. ±0.01 mm The accuracy of a measuring device can be quickly assessed and improved by a calibration process most times.
- Repeatability: It shows how agreeable multiple measurements are on the same object, with the same instrument, and by the same appraiser. It indicates the variation coming from the measuring device itself. Repeatability is often presented in six standard deviations (or 6σ) of measurement data.
- Reproducibility: It is the variation from different human appraisers, e.g. A, B, and C. Reproducibility tests are conducted on the same object using the same instrument, but by different appraisers. Reproducibility, as a variability, is presented in terms of a standard deviation (σ).

Figure 7.2 (a) shows the accuracy and repeatability of a measurement instrument, and Figure 7.2 (b) displays the reproducibility related to the people, procedure, etc.

There are additional types of measurement errors. For example, a measurement's *resolution* is about the ability to detect minor differences between increments of measure. A rule of thumb for resolution is called the "10 to 1 rule," meaning the smallest measurement increment should be 10 times smaller than the requirement for the object being measured. For example, if an object being measured is to the nearest 0.1 mm, then the measurement device should be able to read to the nearest 0.01 mm.

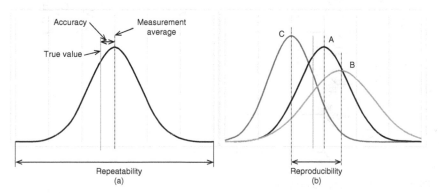

**Figure 7.2** Accuracy, repeatability, and reproducibility of measurement.

*Stability* is another consistency characteristic of a measuring device. Stability means that measurement instrument readings do not drift over time because of environmental factors, such as temperature, vibration, or others. For instance, certain measurement instruments, such as a coordinate measuring machine (often called CMM), should be placed in a temperature-controlled room to ensure stability.

Measurement *linearity* is the accuracy of measurements of different sizes of objects over the full range of a device, which is often presented as a percentage, table, or graphic of the full measurement range. A linearity analysis is often conducted jointly with a bias study, which shows how close measurements are to the reference values in the specific range.

Resolution, stability, and linearity of an instrument are design specifications. They should be addressed when selecting an instrument, but are not often a concern during use. Thus, an MSA normally does not include them.

**Process of MSA**

An MSA is normally on the repeatability and reproducibility or the data spread-related errors of a given measurement system. Therefore, an MSA is often called a "gauge repeatability and reproducibility" test (Gauge R&R or GR&R in short). Figure 7.3 shows a typical MSA assessment process flow.

In MSA processes, method, procedure, and device are the key influencing factors to repeatability variation. Human appraisers can be the primary source of reproducibility variation, while the procedure and method affect

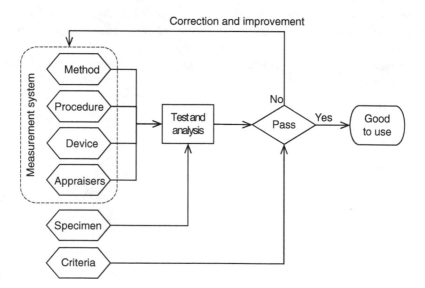

**Figure 7.3** MSA process flow.

reproducibility as well, because of the appraisers using the given procedure and method. A hidden factor that also affects the repeatability and reproducibility is the training and skills of the appraisers. An MSA provides an overall assessment of a measurement system, while finding influence factors would require additional effort.

The results of an MSA should meet the predefined criteria to qualify the measurement system to measure a product or service. If an MSA test fails to meet the passing criteria, a root cause analysis should be conducted, and follow-up correction implemented. Then, an MSA must be conducted again until meeting the criteria. Otherwise, the measurement system is disqualified for the intended measurement tasks.

As an assurance to quality measurements, an MSA should be conducted regularly. An MSA should also be conducted when:

- A measurement system has undergone a change, e.g. equipment, operator, process, etc.
- An investigation shows the root cause of an issue is related to a measurement system.
- The validity of the results generated by a measurement system is called into question.

MSA has been implemented across many industries, including hospitals (Erdmann et al. 2010). Some organizations have specific MSA standards. For example, ASTM (formerly known as American Society for Testing and Materials) has E2782 Standard Guide for Measurement system analysis (ASTM 2017); AIAG has a guideline in MSA (AIAG 2010); and ISO 10012 specifies generic requirements and provides guidance for measurement processes (ISO 2015a).

### 7.1.2 Analysis in MSA

#### Gauge R&R Analysis

A typical MSA focuses on the repeatability and reproducibility of a measurement system, as shown in Figure 7.4. Other technical specifications of measurement system variation can be checked via a calibration process of a measurement device.

The total variation ($\sigma_{total}^2$) of measurements comes from two sources: test objects and a measurement system. Their relationship can be represented as:

$$\sigma_{total}^2 = \sigma_{object}^2 + \sigma_{MS}^2$$

$$\sigma_{MS}^2 = \sigma_{repeatability}^2 + \sigma_{reproducibility}^2$$

where $\sigma_{object}^2$ is the true variance of objects; $\sigma_{MS}^2$ is the variance of a measurement system; $\sigma_{repeatability}^2$ and $\sigma_{reproducibility}^2$ are the variances of repeatability and reproducibility, respectively.

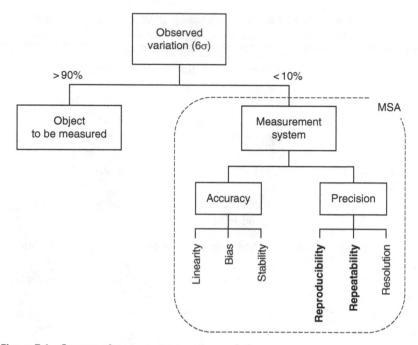

**Figure 7.4**  Sources of measurement system variation.

The repeatability and reproducibility of a measurement system may be tested in one setting. Figure 7.5 shows two scenarios: (a) using shared or crossed samples for non-altered or non-destructive situations, and (b) using unshared or nested samples for altered or destructive tests. Nested tests have higher requirements, with a critical assumption of homogeneous samples, or very small sample-to-sample variation. In a GR&R study, one needs to make sure the same measuring device is used. If possible, MSA appraisers should be operators who would regularly use the measurement device. The result of a GR&R is often presented in six standard deviation ($6\sigma$).

Sampling is important for a robust GR&R. The samples should be representative, e.g. randomly selected from the whole. In addition, multiple samples, say 10 parts depending on part availability, are needed. Each sample is measured twice for reliable readings and validation.

With test data, the manual calculation of repeatability and reproducibility is not difficult. Most quality control books have the equations, which can be used to develop MS Excel tools for GR&R. General statistical software, such as Minitab (Figure 7.6), has GR&R study functionality. There are a few dedicated packages of software on gauge studies.

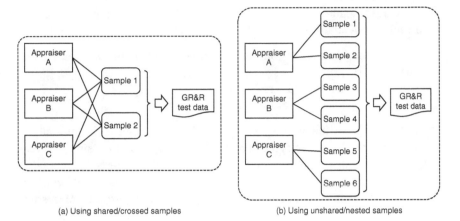

(a) Using shared/crossed samples        (b) Using unshared/nested samples

**Figure 7.5**   Process flows of GR&R studies.

## Attribute GR&R

Many characteristics and features of a product or service are attribute (qualitative or discrete) data when they are counted. Unlike variable data measurements, an attribute measurement shows whether the part is accepted or rejected. This may be a gray area between pass and fail, where wrong decisions can be made.

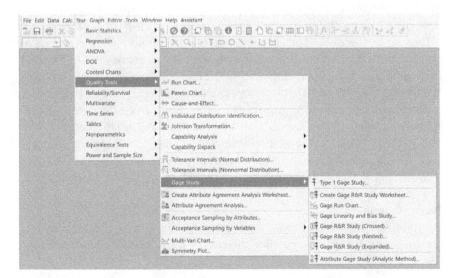

**Figure 7.6**   GR&R study functions of Minitab. *Source*: Minitab Inc.

Following the same principles of variable GR&R, similar rules can be seen for an attribute MSA study:

1.  Equal number of samples from each category (number of good = number of bad)
2.  Multiple, e.g. three, appraisers
3.  Each sample is measured twice

An attribute MSA follows the same principles, but uses a different analysis method, called an attribute agreement study. Such a study may simply reveal agreement percentages in repeatability and reproducibility tests. In general, a resulting agreement above 90% is adequate, and 100% often is a goal, while an 80–90% may be conditional. Using agreement percentage in attribute MSA studies is straightforward. For example, an attribute MSA was used for a university library assessment (Murphy et al. 2009).

The kappa ($\kappa$) statistic, or a coefficient of concordance, is often used to show agreements from contingency tables, obtained from paired samples on pass/fail. The pairs can be between appraisers, an appraiser to a standard, or within an appraiser. The reproducibility agreement between appraisers is also called "effectiveness" in an attribute MSA. $\kappa = 1$ indicates a perfect agreement, while $\kappa = 0$ means no agreement. A practice requires $\kappa > 0.95$ (Rich and Kluse 2018). In theory, $\kappa$ can be negative, but it rarely occurs because a negative $\kappa$ means that the test result is worse than a random chance. Note, for an attribute GR&R of good/bad, or pass/fail, the $\kappa$ does not consider the magnitude of the agreement. Furthermore, an attribute GR&R needs a large sample size, e.g. > 20, to have a reliable conclusion. Figure 7.7 shows a summary page of a GR&R report.

In addition, there is a Kendall's coefficient of concordance, which is an indicator for ranking data with three or more categories, as it accounts for the difference in magnitudes among appraising results. The higher the value is, between 0 and 1, the better the agreement between appraisers themselves, between an appraiser to a standard, or within an appraiser itself. When needed, one may use statistical software for Kendall's coefficient calculation. Attribute and ranking data are common in quality fields. Some studies have used Kendall's coefficient of concordance, for example, the quality assessment of three-dimensional printed surfaces (Fastowicz et al. 2020) and gastrointestinal procedures (Magalhães et al. 2020).

**Passing Criteria**

Interpreting GR&R study results is critical to deciding on the acceptability of a measurement system. The passing criterion of a GR&R study for instrument variation in $6\sigma$ is as follows:

1.  As a common practice, a GR&R result in $6\sigma$ should be less than 10% of the specification tolerance of the products measured. For example, if the tolerance of a part feature is $\pm 0.1$ mm, or a tolerance of 0.2 mm, then the GR&R

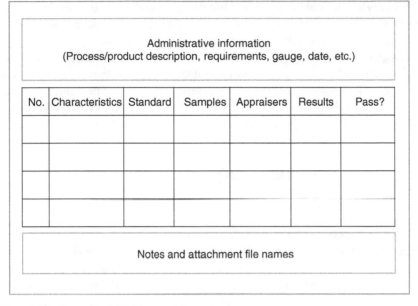

**Figure 7.7** Example of GR&R report format.

of a measurement system should be 0.02 mm or less in $6\sigma$. Accordingly, the measurement system variance ($\sigma^2_{MS}$) is about 1.0% of the total variation ($\sigma^2_{total}$), or $\sigma^2_{total} \approx \sigma^2_{object}$.

2. If a GR&R in $6\sigma$ is at a level of 30% of the specification tolerance of a part, then the variance ($\sigma^2_{MS}$) of the measurement system is about 8.3% of the $\sigma^2_{total}$. This high measurement variation should be unacceptable; and the measurement system and process must be improved.

3. In industrial practices, if a GR&R result is between 10% and 30%, then it may be acceptable based on other factors, such as time or cost to improve and the importance of the quality dimension. Occasionally, the specification tolerance of a product can be challenged. If using a measurement system with such a GR&R variation, a second measurement system is recommended to verify the product quality.

For an attribute GR&R, a passing criterion is often $\kappa > 75\%$, while a $\kappa > 90\%$ represents a desired agreement. The passing criteria particularly for the effectiveness of appraisers may be a higher percentage.

The concept and principle of repeatability and reproducibility also apply to certain methods and procedures. In such cases, relative standard deviation (RSD) in measurements is used, expressed as a percentage:

$$RSD = \frac{s}{|\bar{x}|}\%$$

where $s$ is the standard deviation of test samples, and $|\bar{x}|$ is the absolute value of the mean of test samples. For a method validation, it is expected that RSD < 1%, while for a test, a common practice is that an RSD < 3%.

A high variation of repeatability and reproducibility may occur because of the lack of standard operating procedures, ambiguous instructions, and/or the lack of training for appraisers. If such a root cause can be confirmed, improvement of appraiser training and operating procedure can resolve the high variation issues, without replacing a measurement instrument itself.

## 7.2 Process Capability Study

### 7.2.1 Principle of Process Capability

**Concept of Process Capability**

Process capability is an indicator for how well a process meets design require-ments, and whether it is capable of doing so. Because process capability is a statistical concept and analysis, its prerequisite is having the process in a stable state. As an important assessment and assurance to make quality products, pro-cess capability should be conducted and proved before starting full-scale, nor-mal operations of a production or service. Therefore, a process capability study is another important part of quality planning. For example, APQP schedules process capability tasks during the process design and validation phases. In addition, a process capability study is a tool often used for root cause analysis, for continuous improvement in normal operations.

Two types of information are involved in a process capability study. One is design specifications, often represented with a range, i.e. an upper specific limit (USL) and lower specific limit (LSL), e.g. ± 1.0 mm and + 0.25/−0.05 mm. A product's design determines its USL and LSL, based on customer expectations and product functionality. The other set of information is the actual data col-lected from a manufacturing process or operation. In most cases, it is assumed that these data follow a normal distribution.

The two sets of information can form four situations (see Figure 7.8). A process is fully capable if all actual data are within the specification tolerances, like the situation in Figure 7.8 (1). In this situation, the variation of actual data is smaller than design specification tolerances; the average of actual data is centered in the specifications. Other situations include actual data that fall off-center, and some data spread beyond USL/USL. Figure 7.8 (2), (3), and (4) show these situations.

To quantify a process capability, the common process capability indices (or ratios) are defined $C_p$ and $C_{pk}$ ratios:

$$C_p = \frac{USL - LSL}{6 \times s} \; ; \; C_{pk} = min \left\{ \frac{USL - \bar{x}}{3 \times s}, \frac{\bar{x} - LSL}{3 \times s} \right\}$$

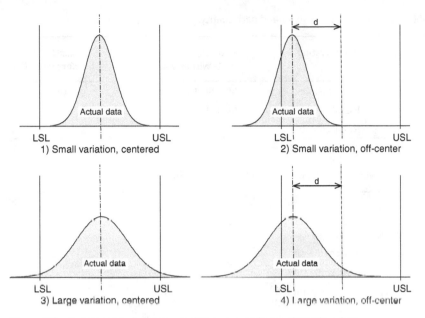

**Figure 7.8** Relationship between design specs and manufacturing data.

where $C_p$ is a simple indicator of the overall process capability, while $C_{pk}$ is a capability indicator that can also express whether the actual data is off the center of a design specification. In the $C_p$ and $C_{pk}$ equations, $s$ and $\bar{x}$ are the standard deviation and average, respectively, of the data.

Both $C_p$ and $C_{pk}$ are dimensionless. Because they are specific to unique tolerances, they may not be used to compare two processes with two different specification tolerances.

### Criteria of Process Capability

A capable process can be viewed as a small variation of actual data compared to design specifications (USL/LSL), and distributed around its center, which indicates that the manufacturing process satisfies the design requirements. In other words, a capable process should have its average $(\bar{x})$ being at least three standard deviations $(s)$ away from the upper and lower specification limits.

To quantify a process capability, the calculated results of $C_p = 1.67$, 1.33, and 1.00 are shown in Table 7.1. The data in the table are true only if the actual data is perfectly centered in specifications (USL/LSL), referring to (1) and (3) in Figure 7.8. Therefore, the parts within their USL/LSL and defect rate PPM (parts per million) are labeled "at most," and "at least," respectively, in the table. If $C_p = 2.00$, the process is in a "$6\sigma$" performance, which has 0.002 defect rate

**Table 7.1** Process capability $C_p$ and part quality.

| $C_p$ | Process rating | At most, parts within USL/LSL | At least, defect rate PPM |
|---|---|---|---|
| 1.67 | Very capable | 99.99994% | 0.57 |
| 1.33 | Passable | 99.994% | 63 |
| 1.00 | Not acceptable | 99.7% | 2,700 |

**Table 7.2** Typical passing criteria of process capability.

| Process status | Capability criteria | |
|---|---|---|
| 1. Capable and centered | $C_p \geq 1.67$ | $C_{pk} = C_p$ |
| 2. Capable and not centered | $C_p \geq 1.67$ | $C_{pk} \geq 1.33$ |
| 3. Not capable and centered | $C_p < 1.33$ | $C_{pk} = C_p$ |
| 4. Not capable and not centered | $C_p < 1.33$ | $C_{pk} < C_p$ |

PPM with centered data. For most products, $C_p = 2.00$ may not be necessary due to high technical requirements and costs.

Based on the equations, $C_{pk} = C_p$ only if actual data are distributed perfectly around the center of specification. In the real world, actual data from any manufacturing process is rarely centered, so it is normal for $C_{pk} < C_p$. If data is far off-center, i.e. d > 3 × s, then $C_{pk}$ is negative. Some six-sigma practitioners assume the off-distance d = 1.5 × s, which may or may not be accurate.

Table 7.2 lists these four situations mentioned, and typical capability passing criteria. Many companies prefer using $C_{pk}$. In most cases, a $C_{pk} \geq 1.67$ is desirable, as it indicates a very good process capability, while a $C_{pk} \geq 1.33$ is often acceptable, if $C_p \geq 1.67$.

Although they are all related to design specifications (USL/LSL), process capability and process control are different. Process control is for both process and status, focusing on removing special-cause variation to improve process capability.

**Discussion of Service Process Capability**

The concept and analysis of process capability studies have been widely used in the manufacturing industries. Considering the uniqueness of services, in theory, one may say that service processes can be evaluated based on the same principle. However, service quality is more related to a customer's feeling and perception than product quality, as discussed in Chapter 1. Notably, the five main dimensions of service quality (Table 1.3), i.e. reliability, responsiveness, assurance, empathy, and tangibility, may be measurable either quantitatively or qualitatively. For example, service responsiveness may be quantitatively measured in minutes, while empathy is measured on a qualitative and subjective scale of 1–5.

In addition, conventional process capability is only applied to stable processes. This requirement implies that it cannot apply for a service or business process, because it tends to be unstable by nature. New process capability methods are needed for real-time, non-normal-distribution processes in service. For example, a combined chart was proposed to display a frequency histogram, with an overlaid cumulative distribution function, estimated from a sample population (Turunen and Watson 2021).

Objectively, there are challenges in this endeavor. A common one is how to make service quality quantitatively measurable for the direct interaction between a service provider and a customer. Additionally, the interaction attribute is affected by uncontrollable or unknown factors; this phenomenon could be an interesting research topic.

Another uniqueness of service quality measurement is related to quality specification. The quality of physical products often falls under a "nominal-is-best" scenario, within a tolerance. Conversely, the quality or customer satisfaction of most service processes is often in one direction, e.g. the shorter the better, or the more the better. For example, the specification tolerance for a patient's waiting time could be USL = 10 min for a doctor office visit. Some services may even not have a fixed USL or LSL, varying with other conditions.

Because of these differences, the measurement and assessment of service capability are more complex than those of manufacturing process capability, and remain a subject of contemporary research. Some studies have focused on using traditional process capability indices in services. One such study showed that service often had a "tolerance zone" instead of a single number limit (Puga-Leal and Pereira 2007). In a tolerance zone, the performance level metric might be perceived differently by customers, or they may have the same level of satisfaction. This observation also introduced the idea that customer satisfaction might be described in a uniform distribution. Another study proposed a new KPI (key performance index or indicator) based on the principle of traditional process indices, and five SERVQUAL (discussed in Chapter 1) categories (Pan et al. 2010).

### 7.2.2 Process Capability Assessment

#### Evaluation of Process Capability

The evaluation of a process capability in manufacturing operations can be conducted in six steps (see Figure 7.9):

1. Select product features and process characteristics
2. Understand passing criteria of $C_p$, $C_{pk}$, etc. to be used

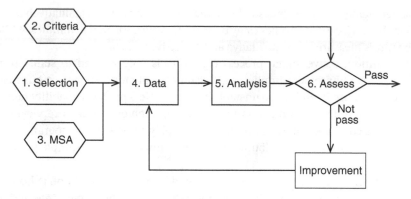

**Figure 7.9** Process flow of process capability assessment.

3. Confirm the successful completion of an MSA
4. Collect data in the planned sample size and frequency
5. Review and analyze data characteristics (pattern, distribution, etc.)
6. Assess process capability, and draw conclusions

To assess process capability, there must be sufficient data within a sampling group or period to calculate the standard deviation of the sample. In a trial run for mass production, one should run a considerable number of parts, such as a minimum of 100 consecutive parts, before calculating $C_p$ and $C_{pk}$. Such a production volume requirement should be specified in a production part approval process (PPAP) package (discussed in Chapter 6), if a part is provided by a supplier.

Process capability ratios are quality indicators, showing process quality status under a certain condition. Process capability ratios can be used as a continuous monitoring index throughout operation time. Process capability is often used as an input to a problem-solving or continuous improvement project (more on this in Chapter 8).

**Discussion of Process Performance**

During early development stages, the aforementioned prerequisite for sample size may not be satisfied. In those cases, $C_p$ and $C_{pk}$ should not be used. Instead, similar indicators, called process performance indices $P_p$ and $P_{pk}$, are recommended by AIAG (AIAG 2006). They are represented as follows:

$$P_p = \frac{USL - LSL}{6 \times \sigma}; \; P_{pk} = \min\left\{ \frac{USL - \bar{x}}{3 \times \sigma}, \frac{\bar{x} - LSL}{3 \times \sigma} \right\}$$

where $\sigma$ is the standard deviation of *all* available samples. The difference between process capability ($C_p$ and $C_{pk}$) and process performance ($P_p$ and $P_{pk}$) is their standard deviations, "*s*" vs. "$\sigma_p$" in the equations. In terms of passing criteria, similar criteria values of 1.67 and 1.33 are often adopted.

Some quality professionals have different viewpoints on $P_p$ and $P_{pk}$ (Kotz and Lovelace 1998; Montgomery 2013). They argue that without knowing whether a process is in statistical control, $P_p$ and $P_{pk}$ cannot correctly express the performance level of that process. A key factor affecting process capability evaluation in development phases is small sample size, or short run in different trials. In such situations, a process evaluation loses its statistical sense. The samples from different trials of development may be in various settings and have different distribution characteristics, e.g. different average and standard deviation values. When all the data is combined, it is difficult to prove data normality. It is also challenging to interpret the meaning of the standard deviation calculated from fundamentally different samples. Therefore, the calculated values of $P_p$ and $P_{pk}$ can be confusing.

Some companies require a process to be stabilized, under control, and of data normality before using $P_p$ and $P_{pk}$. For example, one simple test is to use a data set of at least 25 subgroups, and a subgroup size of at least five measurements, to determine process stability and normality. If the data collected are not statistically stable and/or normally distributed, special causes to variation should be identified and eliminated before proceeding.

One way to estimate process performance is by comparing the collected data to design specifications. If the performance data fall within the specifications, without data on the borderlines, in a test, the test may be deemed successful, for the time being. Otherwise, further analysis is necessary to determine root causes for disagreements, and make any necessary corrections.

In the pharmaceutical industry, the US Food and Drug Administration (FDA) requires a stage appropriate process conformance before a new active pharmaceutical ingredient, or a finished drug product, is released for distribution. The FDA does not specify a minimum number of batches to validate during a manufacturing process, but expects that a manufacturer has a sound rationale for its choices (FDA 2019).

### 7.2.3 Production Tryout

**Tryout Process**

Using short production runs (or pilot runs) is a common practice of tryouts. For a product tryout, one may use temporary tooling and processes. If a tryout serves to test the readiness of manufacturing processes, and to validate the performances for ultimate customers before mass production, it should be in a production-intended setting. Launching a new product directly into a production system without tryouts can be very risky.

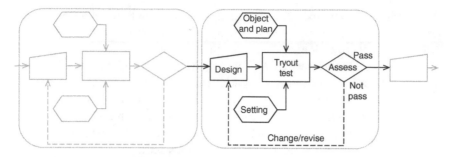

**Figure 7.10**   Process flow of tryout tests.

A tryout document has a specific test objective and plan, and test setting. Tryout results are evaluated based on these predefined criteria, to determine the verification satisfaction. If it passes a tryout, then a given design moves to the next task or stage. Figure 7.10 shows a process flow chart of a tryout test. This generic tryout process flow may apply to service development as well.

The conditions of a production trial include operating in a production facility, using product parts, using production tooling and equipment, operating by production workers, and running within a designed cycle time. Such a validation process is often called a production trial run, which needs significant preparation effort. The issues from previous development phases could prevent a run at the capacity required by the current development phase. These previous issues must be resolved prior to a production trial.

In the middle of development, more and more tryouts are conducted in a virtual environment using computer simulations. Many functions, features, and performance attributes of a product can be tried out virtually for verification. For a manufacturing system, a virtual tryout (often called virtual commissioning) can reach over 90% accuracy to actual situations; meaning relatively minor adjustments can make the real system work. However, the last phase of a product tryout must still be conducted in a real production setting.

### Tryout Preparation

To prepare a production trial, a so-called "preflight checklist" is often developed and used in industry. Table 7.3 shows a readiness checklist as an example. The items in the exemplified checklist are related to the items and functions that a manufacturing production system should be ready to run a trial.

Checklist items are application dependent. One can develop a similar checklist for a specific tryout run. For example, a process capability certification may have several aspects, such as dimensional, structural, and surface quality, while a tooling quality certification may include locating accuracy and repeatability. These detailed items may be checked separately, prior to the final preflight checklist.

**Table 7.3** Sample of preflight checklist for a production trial run

| Administrative | Area/line: | | Date: | Approval: |
|---|---|---|---|---|
| | Responsible: | | Team: | |
| **Category** | **Item** | | **Readiness (Y/N/NA)** | **Requirement and remarks** |
| System/process | Previous issues resolved | | | |
| | System: automation functionality | | | |
| | System: human–machine interface | | | |
| | Cycle time tested | | | |
| | Process capability | | | |
| | Process quality certifications | | | |
| | All operations production intent | | | |
| | ... ... | | | |
| Tooling and equipment | Tooling status | | | |
| | Tooling functionality | | | |
| | Tooling quality certifications | | | |
| | Conveyor (upstream/ downstream) | | | |
| | Internal conveyor/feeder | | | |
| | ... ... | | | |
| Parts | Parts quantity | | | |
| | Parts quality | | | |
| | Parts line side readiness | | | |
| | Parts/material handling | | | |
| | ... ... | | | |
| Manual operations | Operation safety verification | | | |
| | Operational instructions | | | |
| | Operator training | | | |
| | Ergonomic certification | | | |
| | ... ... | | | |

A cross-functional team needs to review these checklists to ensure the items related to their responsibility are completed. If an item on a checklist is not fully ready, it can jeopardize a trial run. Therefore, management approval is needed for such items. After a trial run, the checklist may be revisited to develop any lessons learned, for future tryout preparation.

**Passing Criteria of Tryout**

The passing criteria of a verification production trial are based on the characteristics of the products, processes, and operations. Most of them are specified during design, and/or in the PPAP, if a trial and later production are at a supplier site. It is a good idea to select a few key, rather than all, characteristics for a trial run. These key characteristics might include:

- Product: Quality characteristics, e.g. dimensional and structural.
- Process: Capability characteristics, e.g. automation, $C_p$, and speed.
- Operation: System performance, e.g. cycle time and throughput rate.

**Table 7.4** Example of performance report of a production trial run

| Area/line: | | Date: | | Approval: | |
|---|---|---|---|---|---|
| Responsible: | | Start time: | | No. of units: | |
| Team: | | End time: | | | |
| | Item | Target (final) | Target (this run) | Actual result | Follow-up |
| System | Downtime (min/h) | No | < 5 | | |
| | Cycle time (second) | 55 | 30 | | |
| | Throughput (JPH) | 60.2 | 35 | | |
| | ... ... | | | | |
| Process | Welding quantity | 100% | 100% | | |
| | Welding quality | 100% | 80% | | |
| | Sealing quantity | 100% | 100% | | |
| | Sealing quality | 100% | 70% | | |
| | ... ... | | | | |
| Product | Dimensional quality | 98% | 50% | | |
| | Surface quality | 98% | N/A | | |
| | ... ... | | | | |

A trial run can be conducted during different development phases. If used for the same purpose, each time may have an increasing target value to the 100% target goal in a normal production. For example, a starting passing criterion may be set at 70% of the ultimate targets, and the target in the following tryout could be 85%. This way, the farther along a development process is, the higher the standard becomes. Industry practices show that multiple production trials work well to identify issues and gradually approach a final verification target. Additionally, different production trials can be purposely planned for a specific focus each time. Corrective follow-up plans and actions are necessary for unsatisfactory trial runs.

A formal trial run report, with the open issues documented, is usually required. Table 7.4 shows a report example of a production trial for a vehicle body subassembly.

## 7.3 Change Management in Development

### 7.3.1 Process of Change Management

#### Significance of Change Management

Change management is a common task in the business world, which can improve performance, increase profits, enhance competitive advantages, etc. This section focuses on the technical changes in product and service development, rather than on the change or transformation in organizational business or operations in general.

The development of a new product or service can be complex, and take months or years to complete. Because of the nature of development, changes during development are inevitable, to meet a variety of challenges and driving factors. Therefore, change management is a part of development and quality planning. For example, change information is part of and an output of Phase 2 – product design and development of APQP (discussed in Chapter 1). Over the life cycle of a product or service, changes can also be common, and their effective management is important.

The driving factors for change include revised objectives, updated requirements, novel ideas, issues found in trials, design reviews, and adoption of new technologies. Changes normally require analysis, evaluation, and project management. Managing changes affects timely deliverables and customer satisfaction, and is a large part of a successful product or service. Change management correspondingly plays a significant role in leveraging design efficiency, shortening the time span of development, and reducing resources requirement.

Change management itself is a process throughout the entire life of a product or service, from conceptualization to routine operations, through maintenance, and ultimately until obsolescence. Based on the nature and timing of changes, they can be viewed by two types: changes during development, and changes during a regular operation or production. This book focuses on the former, as the latter is often more closely related to continuous improvement.

Design changes are common during development. They can be adaptive (relatively small, gradual, or iterative), transformational (or large in scale, or with an intense impact), or anything inbetween. Many changes, particularly late ones, come with potential risks. To assess the likelihood of risks associated with changes, one may use a risk-based management process (reference to Chapter 2) to evaluate and justify a given change request. Through acknowledging the impacts and risks of a proposed change, project management understands the significance of the change, balances tradeoffs, and decides whether to approve or disapprove a change request.

### Management Process for Change

Change management is the process of guiding a change proposal to resolution. The process of design change management involves six primary tasks (see Figure 7.11) in product development, with a central focus on the first four. Some change management systems may have fewer tasks by combining two tasks, or have more tasks by going into further detail than others. This process also applies to other areas, such as service and manufacturing process planning.

1. <u>Change identification.</u> First, a design team considers a change and preliminarily identifies its benefits/risks and feasibility, in terms of technical, financial, and timing considerations.
2. <u>Change request initiation.</u> Based on the preliminary analysis, the design team agrees on the necessity and value of the change, and formally initiates a change notice (CN) or request for change (RFC) into a business system.

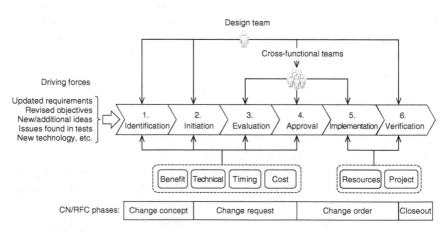

**Figure 7.11** Process flow of design change management.

3. <u>Cross-functional evaluation.</u> Cross-functional teams review the CN/RFC on the benefits, feasibility, cost, impacts, risks, and quality of other functions. All the related parties have their inputs into the business system. It is common that some parties disagree with the change, or raise a major concern. The design team communicates with concerned parties, and comes out with feasible resolutions.

4. <u>Change approval.</u> Based on the inputs from all related parties, the change authority committee approves or disapproves the CN/RFC. The committee should have the authority to adjust the timing and budget, on other aspects of the development project to account for the approved change.

5. <u>Change implementation.</u> If the CN/RFC is approved, the design team shall plan and coordinate the implementation efforts, including design revision, testing, and documentation update.

6. <u>Change verification.</u> After implementation, the design team shall verify the actual status of, and finally close out, the CN/RFC as appropriate.

A CN/RFC project has four phases: concept, request, order, and closeout. These four phases are also included in Figure 7.11, showing the relative relationship between the six tasks and four phases. The four-phase expression is often used in a change management system to track CN/RFCs.

**Main Tasks**

During a change management process, a design team plays a central role in communication. A proposed change likely brings up additional workload for others, and affects project timing and expenditures. The tasks of change management, depending on the type and significance of a proposed change, normally include various analyses, evaluations, and approvals.

There are dedicated software and software modules in product life-cycle management (PLM) software for change management. One challenge to change management is the necessity of prompt responses from other concerned departments. A change management system should have functions to enhance review/feedback responsiveness.

For important products, such as medical devices and drugs, change management should not only follow the aforementioned process but also conduct intrinsic, specific tasks. The FDA suggests that a change management system should include risk management, market authorization, expert evaluation, and confirmation, as appropriate for the stage of a given life cycle:

"(a) Quality risk management should be utilized to evaluate proposed changes. The level of effort and formality of the evaluation should be commensurate with the level of risk.

(b) Proposed changes should be evaluated relative to the marketing authorization, including design space, where established, and/or current product and process understanding. There should be an assessment to determine whether a change to the regulatory filing is required under regional requirements. As stated in ICH Q8, working within the design space is not considered a change (from a regulatory filing perspective). However, from a pharmaceutical quality system standpoint, all changes should be evaluated by a company's change management system.

(c) Proposed changes should be evaluated by expert teams contributing the appropriate expertise and knowledge from relevant areas (e.g., Pharmaceutical Development, Manufacturing, Quality, Regulatory Affairs, and Medical) to ensure the change is technically justified. Prospective evaluation criteria for a proposed change should be set.

(d) After implementation, an evaluation of the change should be undertaken to confirm the change objectives were achieved and that there was no deleterious impact on product quality." *(FDA 2009)*

### 7.3.2   Considerations in Change Management

#### Change Evaluation

A core element of design change management is an impact-focused and risk-based change evaluation. The evaluation should cover all related factors, potential risks, and affected parties. The influencing factors to a design change may be categorized into six groups, and evaluated during a review and approval process. Table 7.5 is an example of an overall evaluation matrix with the six factors of a CN/RFC, to help evaluate and decide if the change is warranted.

In addition to the overall evaluation, each affected department should assess the impacts on their own interests if a CN/RFC is approved. For example, a manufacturing operation should evaluate CN/RFC-associated physical modifications on the aspects of equipment (machine), process (method), personnel (work force), and material. Customer services and sales departments may need to have respective changes as well.

To manage design changes effectively, one can categorize potential changes into three levels.

1. Minor changes or intra-department changes. This type of change is minor and has little influence or risk on other functional departments. Thus, these minor changes can be quickly handled within the design team. A CN/RFC may still be needed for documentation.
2. Limited impact/risky changes. This type of change affects a downstream development process and the final product, or implies noticeable risks, but is within management's technical and financial tolerances. Such a change

**Table 7.5**  Example of evaluation form for design change proposal

| Change type: | Design: | | Date: | | CN/RFC #: | |
|---|---|---|---|---|---|---|
| **Change reason:** | ⊗ Safety  ◯ Quality  ◯ Timing  ◯ Function  ◯ Performance  ◯ <br> other | | | | | |
| **Explanation/ importance:** | **Structural improvement, critical product characteristic** | | | | | |
| **Affected department:** | ⊗ Design ⊗ Quality ⊗ Manuf./Op.  ◯ Fin ⊗ IT ⊗ Supplier  ◯ <br> Other | | | | | |

| Factors: | Very "+" (5) | (4) | Neutral (3) | (2) | Very "-" (1) | Remarks |
|---|---|---|---|---|---|---|
| 1. Design feasibility | | 4 | | | | |
| 2. Not affecting other functions | 5 | | | | | No |
| 2. Manufacturing/ operational | | | | 2 | | Major changes in process |
| 3. No additional cost | | | 3 | | | Minor cost change |
| 4. Project timing/ workload | | | | 2 | | Conflicting with test scheduled |
| 5. Improved quality | 5 | | | | | |
| 6. Support from IT, supplier, etc. | | 4 | | | | |
| Additional issues | N/A | | | | | |
| Unweighted sum and decision | Total score = 25 (out of 30); approved | | | | | |

needs to go through a change management system, and most likely will pass the approval process without major concerns.

3. <u>Major changes</u> with high risks. This type of change significantly affects project timing and budget, or has major technical challenges. Such a change needs thorough review in a change management system, has related department support, and even goes to senior management before an approval/disapproval decision is made.

Risk management approaches, including process analysis, risk evaluation, and risk treatment (discussed in Chapter 2), should be used for major changes. Risk analysis and impact evaluation can come out of quantitative technical and financial assessment reports.

**Discussion of Changes**

The principle and process of change management are straightforward, yet challenging, because of the complexity of business systems. The challenges include:

- Difficulties to keep CNs/RFCs under control due to diverse stakeholders and characteristics.
- Effectiveness and efficiency in implementing change management.
- Data management to ensure only the current documentation available to all functions and departments.
- Communication, processes, and data structure in change management. (Wilberg 2015; Storbjerg et al. 2015; Schuh et al. 2017; Lakymenko et al. 2020)

From a finance and project timing perspective, the fewer and earlier the design changes are, the better they are, because they have relatively small impacts in a product or service development. There is always room to reduce the number of CNs/RFCs, and drive them to be considered in early development phases, as illustrated in Figure 7.12. In this figure, the quantity and timing of changes made by company B should be improved to reach the better performance level of company A.

For design and other development phases, there are "freeze" dates, which are the deadlines for change proposals. After a freeze date, it would be very difficult

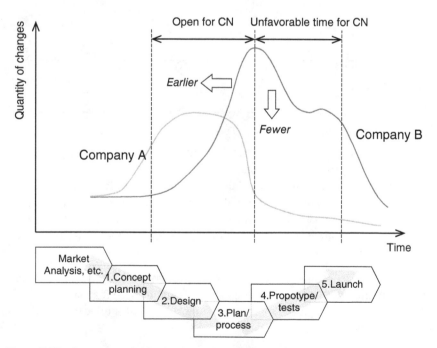

**Figure 7.12** Improvement of quantity and timing of design changes.

to adopt a design change due to project timing. For example, a design is already released for a prototype build. However, if a late change is critical to the customer, then the project management team has to implement it, regardless of cost.

A proposed change can have non-technical obstacles (e.g. unwillingness to a change from the non-proposing departments). Therefore, a change initiator should have a firm understanding of the impacts of the change, and can aptly address the obstacles holistically.

For many changes, the evaluation is on their direct impacts, e.g. using Table 7.5 for possible main impacts. However, the implications of a change may not be fully understood, particularly for the indirect influences and long-term "side effects" due to limited resources, at the evaluation time. Therefore, monitoring and post-change review of the adopted changes are necessary until their effects are well managed.

### 7.3.3 Advancement of Change Management

Change management is not only an integral part of business development but also an applied research subject. From an application demand perspective, researchers in various disciplines have been working on improvement of, novel approaches to automation in, optimization for, and nuance of processes involved in change management. We will now look at a few discussion examples on the various studies in change management advancement.

One case study on the investigation of polypharmacy management in nine European countries (McIntosh et al. 2018) highlighted the significance of change management and theory-based implementation strategies, and provided examples of polypharmacy management initiatives.

Some changes can be late after implementation. In another study, Comuzzi and Parhizkar (2017) proposed a methodology for controlled change management for a post-implementation modification. They defined a unique methodology, as a set of artifacts, to support the different phases of change impact analysis, and described a set of mechanisms to assess the impact of a proposed change.

A further case study was conducted on a major automotive company, based on a survey of 46 people and interviews of 12 people (Knackstedt 2017). The author found that the change management process was in place and was followed during business operations. They also found, from employees' feedback, that the management considered only a portion of the total engineering input available when deliberating a given change. The study concluded that the quality of information entered during review was poor, and found a lack of communication throughout the change management process.

In the software engineering field, gaps have been found in change identification, change analysis, and change cost estimation. Jayatilleke and Lai (2018) identified the importance of having a full-scale model to increase the efficiency of managing change with better accuracy.

A literature review on change management showed that limited work had been done on pre-change in people-oriented change reduction measures, in a change stage regarding the organizational issues and implementation of changes, and in post-change on the impact of changes on quality and manufacturing and post-manufacturing stage (Hamraz et al. 2013). These issues may imply research opportunities to fill these gaps, to make change management grow to become more effective.

Researchers also study new methods and frameworks for change management. One study proposed an information table with a systematic flowchart, to automate engineering change management (Zheng et al. 2019). The authors of this study also introduced a co-occurrence-based design structure matrix, and three-way based cost-sensitive learning approach for change prediction. Other researchers have used computer simulation to study current processes, and several possible evolutionary scenarios (Lellis et al. 2018). Based on their simulations, they proposed a framework that allows getting quantifiable feedback on the evolutionary scenarios of the analyzed process with minimum effort and time.

Another arena of advancement in change management is the spectrum of challenges in developing a highly user-friendly, cloud-based platform designed to improve process effectiveness of change management and to prevent some of the newly discovered issues mentioned.

## 7.4 Quality System Auditing

### 7.4.1 Roles and Processes of Quality Auditing

#### Purpose of a Quality Audit

A quality system, including its mission statement, procedures, and standards, should be established in a business. One way to ensure that operations (and the people involved) are fully in line with the mission statement, and following procedures and standards, is employment of a regular audit. A quality system audit is a formal monitoring and examination process, conducted by either internal or external quality auditors. An audit is also a key element in the ISO quality system standard ISO 9001.

The purpose of a quality system audit is to assess not only products or services but also their development and operation processes. A quality system audit normally focuses on:

- Operations that comply with governing source documentation, such as corporate directives, standards, and government regulations.
- Actual effectiveness of QMS procedures and processes, and their results.
- Identification of the best practices in a given organization or industry, and identification of areas for improvement.

A quality system audit can reveal the degree of conformance of a system or organization with governing documentation in following the defined quality processes and in meeting conformance standards. In addition to assessment, the other purpose of an audit is to find improvement opportunities of a system and guide follow-up actions.

For example, a quality system is essential to the food, medicine, and medical device industries. The FDA states on the management control of inspections: "Management reviews and quality audits are a foundation of a good quality system. Assure that the manufacturer has written procedures for conducting management reviews and quality audits and there are defined intervals for when they should occur" (FDA 2014a).

As such, some disciplines have government regulations or requirements on a quality system audit. The FDA requires quality auditing to be performed as part of its quality system regulation for medical devices. The FDA considers that auditing is an assessment of performance indicators that can monitor the effectiveness of processes within the pharmaceutical quality system (FDA 2009). The Code of Federal Regulations (CFR) Title 21, Part 820 states:

> "Each manufacturer shall establish procedures for quality audits and conduct such audits to assure that the quality system is in compliance with the established quality system requirements and to determine the effectiveness of the quality system. Quality audits shall be conducted by individuals who do not have direct responsibility for the matters being audited. Corrective action(s), including a reaudit of deficient matters, shall be taken when necessary. A report of the results of each quality audit, and reaudit(s) where taken, shall be made and such reports shall be reviewed by management having responsibility for the matters audited. The dates and results of quality audits and reaudits shall be documented." *(CFR 2020)*

### Process of Quality Audit

Some other types of checks, such as a specific document review and a routine inspection of a final product, are often called an audit. However, a quality system audit should be a systematic and comprehensive assessment (and may include the quality performance of a product and service). A quality system auditing process is shown in Figure 7.13. The figure also shows the individual responsibilities and teamwork between auditors and auditees. An audit on specific areas may be carried out repeatedly, on a predefined time interval.

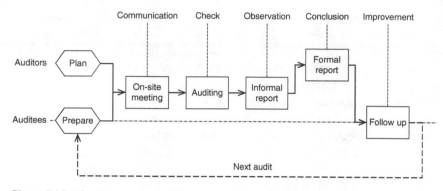

**Figure 7.13** Quality system auditing process flow.

After auditing, auditors create an audit report of findings and conclusions, which are discussed with auditee department management. The audit conclusion regards the conformance of the audited subjects to the applicable standard or audit requirements. An audit conclusion may include statements and comments regarding the suitability of a process to achieve objectives, and suggestions for the possibilities of improvement. The issues and nonconformities in the findings are often called corrective action requests (CARs) as a formal nomenclature, which is a value-added to business improvement. Each CAR needs a follow-up; it is the auditee management's responsibility to assign responsible people and establish a deadline for completion. Some CARs need a follow-up audit.

ISO 9001 standard provides requirements for planning and implementing a QMS (ISO 2015b). A quality system audit can follow ISO 9001 and ISO 19011 standards. ISO 19011 is an international standard to guide auditing management systems. This standard includes the principles of auditing, managing an audit program, and conducting management system audits. ISO 19011 includes guidance for the evaluation of competence of individuals involved in the auditing process (ISO 2018). ISO 9001 and ISO 19011 standards are not specific to an industry, and can apply to organizations of any size.

One common audit practice is called a layered process audit (LPA). The principle of an LPA is that quality assurance is not only for a final product or service but also for the processes of realizing the product or service. As the name indicates, the "layers" of work personnel check different levels of details and frequencies (see Figure 7.14). Team leaders do checklist-based audits of the work cells daily. At the levels of frontline supervisors and area managers, they check fewer critical processes based on predefined metrics. As an LPA focuses on processes, it can provide good quality assurance as in a nature of prevention, and perhaps relegate the evaluation of more critical processes to appropriate managerial tiers.

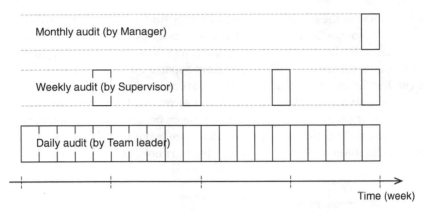

**Figure 7.14** Illustration of an LPA.

### 7.4.2 Types of Quality Audit and Preparation

#### Types of Quality Audit

There are several types of quality audits, which serve specific purposes. Three common types are:

- Systemaudit: A system audit is an overall assessment of a QMS to address who, what, where, when, and how the quality system is used for a given business operation. A system audit verifies whether a quality system has been developed completely, documented appropriately, and implemented effectively. This section of the text is mainly about this type of audit.
- Special audit: This type of audit is on a particular subject, such as a product, process, production, supplier, safety, or facilities. When auditing a process, for example, focus can be on its plan, procedures, work instructions, and standards, and how people follow them. The LPA discussed earlier is a process audit. For a product, an audit can be a type of quality inspection on key characteristics of a product. Because of the inherent connection between a product and process, an audit should address both simultaneously. Internal special auditing can be planned as a regular tool for quality assurance.
- Compliance audit: This is a special audit that compares and contrasts written source documentation to prove (or disprove) compliance with that source documentation. A compliance audit can be on a system, product, or environmental function, against a particular regulation or standard. A compliance audit is often conducted by third-party or external auditors. After a successful compliance audit, a certificate, e.g. ISO 9001 certification, is issued by a certifying authority.

An auditor can be either internal or external. An internal audit, e.g. LPA, is called a first-party audit, conducted by trained internal staff. Internal auditors act on behalf of the company, and are normally from a different department or site. It is also worth mentioning that a self-audit, by an auditor in the same department, is a learning tool. A self-audit can be for a formal audit preparation, for checking the status of and improvement opportunities for a business department. Key for an internal audit's success is to ensure objectivity. External audits can be second-party or third-party audits. A second-party audit is conducted by the customer (or by a contracted organization on behalf of the customer), and a third-party audit is through a professional audit agency. In the case of a third-party audit, the auditors normally check the existing results, rather than technical details.

**Quality Audit Preparation**

In many cases, audit preparation work is planned around an external certification audit, and scheduled throughout the year. To be ready for a quality system audit, the following key elements should be in place:

1. Measurable goals
2. Established QMS, including procedures and standards
3. Employee training on the QMS
4. Implementation of the QMS, with evidence (documents)
5. A computer system available for records and follow-ups (not always required)

To prepare an audit, a checklist or *aide-mémoire* is often developed and used. Table 7.6 shows a generic sample of a preparation checklist. The items on a preparation checklist are based on the business in question, and the overall purpose of the audit. To be practical and specific, one to two pages of a preparation checklist are recommended for each area or team.

The conclusion column in Table 7.6 shows one of three states for each item:

- OK: Meeting all the requirements
- OFI: Opportunity for improvement, a minor issue
- NC: Non-conformance, a major issue

To prepare a formal third-party audit, it is common to give employees refresher training and conduct an informal internal audit in advance. Such a practice can be effective in providing a solid assessment, finding nonconformities to improve, and leading to a better likelihood of passing a formal audit.

### 7.4.3 Considerations in Quality Auditing

**Sampling in Quality Audit**

Because of the size and scope of a QMS, it is often impractical to check and validate each item in a short period, given finite resources. Sampling is a process to obtain and evaluate the evidence of some items and characteristics for a reliable

**Table 7.6** Sample of preflight checklist for quality system audit

| Area/team: | Prepared by: | Date: | |
|---|---|---|---|
| Item | Reference | Evidence | Conclusion |
| Quality vision | | | |
| Quality mission statement | | | |
| Mid-term and annual objectives | | | |
| Pertinent regulations | | | |
| Pertinent corporate policies | | | |
| Catchball process | | | |
| ... ... | | | |
| Accessibility to QMS system | | | |
| Safety requirements | | | |
| Operations processes | | | |
| Quality related work and files | | | |
| Design reviews and records | | | |
| Conformance data and records | | | |
| ... ... | | | |
| Past audit reports and open CARs | | | |
| Continuous improvement | | | |
| Training records | | | |
| ... ... | | | |

audit conclusion. There are two key elements in a sampling plan: sample size and sampling method. Both elements affect the reliability of audit results, individually and in concert with one another. Generally, the larger a sample size is, the more accurately those samples represent the actual situation.

For an external audit, probability sampling methods should be used for a better representation. Common methods include simple sampling (picking any element with equal possibility or fully random) and systematic sampling (e.g. picking every fourth from a complete set). Fully random sampling may be the best way. Then, auditors may draw a conclusion based on the theory of inferential statistics. The reliability or the level of confidence of a conclusion from sample data depends on sample size. For an internal audit, random sampling is rarely used, as management may have a specific purpose and consideration, e.g. on previously identified areas, for conducting an audit to find improvement opportunities. This is called judgment sampling, which does not provide a statistical estimate (or quantification) of the audit findings.

There are publications and regulations for audit sampling, e.g. the FDA Guide to Inspections of Quality Systems presents tables for selecting sample size using statistical methods. If the total number of records is manageable (30 or less), you may choose to review all the records (FDA 2014b). Other publications include those by Walfish (2007) and Murdock (2019).

It is important to conduct an audit on both products and processes. Particularly for mass production or a large variety of products, an audit is often less effective in revealing an accurate status, because of the sheer volume of products. In such cases, an audit should focus more on the processes, to ensure the quality of products via sufficient process control. The audit results of products and corresponding processes should support each other.

As previously discussed, the nonconformities found in an audit can be documented as a CAR, for follow-up and improvement. The number of CARs may or may not precisely reflect the actual gravity of a situation, nor reveal all key issues. Without auditing 100% of target features and documents, the conclusion of a quality system audit is only an overall estimation. This argument may lead to the following questions on the effectiveness of audits.

**Effectiveness of Audits**

Quality system auditing has been a common practice in industries for over three decades. A quality system audit differs from a financial accounting audit. An accounting audit is often mandatory, designed to examine a company's financial statements, while a quality system audit does not bear such a responsibility. The driving forces for quality system audits include:

- Mandated by government regulatory requirements
- Required by customer companies or as part of contract
- To gain a credential of QMS, e.g. ISO 9001 certification, by a third party
- To assess operational status and improvement opportunities

Quality system audits can help find quality issues and improve the quality of many companies. However, quality system audits incur costs, and may be a distraction from regular work. Audit cost is considered as a type of good quality cost, as an audit is part of inspection and prevention. In the meantime, audits can help prevent quality issues, and reduce the cost of poor quality (discussed in Chapter 1).

It is interesting to study and compare the two types of quality costs for an audit practice. The costs and time invested on quality system audits can be calculated, while the benefits and improvement opportunities identified from audits can be assessed as well. Therefore, evaluating cost-effectiveness is plausible. There are some studies on this topic, such as on influencing factors (Salehi et al. 2019; Guliyev et al. 2019), methods (Amiruddin 2018; Refaat and El-Henawy 2019), and cost analysis (Donatella et al. 2018; Reid et al. 2019).

Some studies recognized that there was a disagreement on whether compliance with continuous improvement proviso in most quality standards was practically effective. One study indicated that compliance was generally staff work, and that improvement required proactive management support and enthusiasm (Lahidji and Tucker 2016). It suggested operationalizing the promise of continuous improvement, not just passing an external audit, to enhance market competitiveness, and achieve true excellence.

In terms of lean manufacturing principles, it is controversial whether a quality system audit adds value to customers. The cost associated with quality system audits is not visible to the customers, so they may not appreciate the cost (and time) spent on those audits. The value to customers of quality system audits, i.e. audit cost and associated quality improvement, should be addressed for every unique business case.

An additional consideration to quality system auditing is the necessity and suitability of virtual (or remote) auditing. Due to safety, financial, and logistical constraints, e.g. a pandemic situation or extreme weather, an audit may be tailored to run in a virtual format, preventing cancelation. Remote auditing may offer greater flexibility and yield better stakeholder participation. However, an additional planning effort is needed for remote work, such as understanding unique items, navigating technological issues, and dealing with limitations. ISO 9001 has some guidance on remote auditing (ISO 2020).

## Summary

### Measurement System Analysis

1 Quality is not manageable without measurement.
2 MSA is a process and method for evaluating the measurement errors and qualifying a measurement system.
3 Repeatability and reproducibility (R&R) are the most important characteristics of a measurement system.
4 There are standardized requirements, processes, and criteria for MSA.
5 MSA can be either quantitative or qualitative in character. Both follow the same principle, but have different analyses and criteria.
6 Quantitative MSA passing criteria is that the variation in $6\sigma$ should be less than 10% of the specification tolerance of the targets. If an MSA variation is between 10% and 30%, it may be acceptable considering other factors.

## Process Capability Study

**7** Process capability is an indicator to show how well a manufacturing process meets with design requirements of parts.

**8** There are two common process capability indicators: $C_p$ and $C_{pk}$, while the $C_{pk}$ includes actual off-centered situations.

**9** A common, decent passing criterion of process capability is $C_{pk} \geq 1.67$.

**10** During development phases, similar indicators called process performance $P_p$ and $P_{pk}$ may be used, only if a process is stable, and its data has a good sample size.

**11** For a low-volume tryout, process capability and quality may be assessed directly against target specifications.

## Change Management in Development

**12** Technical changes during development are inevitable. Change management is the process of guiding and handling a change proposal to resolution.

**13** Change management process comprises six steps: identification, initiation, evaluation, approval, implementation, and verification.

**14** A proposed change can be evaluated to three levels of impact: minor, limited, and major, based on the risks, technical issues, financial implications, etc.

**15** A proposed change needs cross-functional team reviews, and management approvals, for potential impacts on the multiple aspects of a product or service.

## Quality System Auditing

**16** A quality system audit assesses products or services and their related processes on QMS aspects.

**17** Quality system auditing processes include on-site meeting, auditing, informal reporting, formal reporting, and follow-up.

**18** An LPA is a quality assurance that involves multiple levels of a workforce, with different audit responsibilities at each layer of the workforce.

**19** Common types of quality audits include system audits, special audits, and compliance audits. In terms of auditors, an audit can be a first-party (internal), second-party (customer company), or third-party (independent agency) audit.

**20** Reviewing the requirements, documentation, and using a preparation checklist can help prepare for a quality system audit.

**21** Sampling is a key factor for quality audits. External audits often use random or probability sampling.

# Exercises

## Review Questions

1 Describe how to define the overall capability of a measurement system.
2 Distinguish between the R&R of a measurement instrument.
3 Review the components and sources of measurement variation.
4 Explain the MSA for attribute data.
5 Explain the passing criteria of MSA.
6 Discuss the factors for USL and LSL determination
7 Discuss the assumptions of using $C_p$ and $C_{pk}$.
8 Review the principles of $C_p$ and $C_{pk}$ with an example.
9 Discuss the passing criteria of $C_p$ and $C_{pk}$ with justification for your claims.
10 Discuss the relation between $C_p$ and $C_{pk}$.
11 Review the tryout process and passing considerations.
12 Review the concept and application of $P_p$ and $P_{pk}$ with an example.
13 List likely reasons for design changes during development of a product or service.
14 Discuss the change management process with an example from an industry.
15 Use an example to discuss the impacts on other aspects of a product or service from a design change.
16 Explain how to evaluate the effects of a proposed change in change management.
17 Use an example to show a common issue in change management practice.
18 Explain the process of quality system audits with an example.
19 Explain the first-, second-, and third-party quality system audits with examples.
20 Review the pros and cons of the different sampling methods in quality system audits.

## Mini-project Topics

1 Justify why the MSA variation should be less than 10% of the total variation, in terms of standard deviation.
2 Search and review an example of attribute MSA application.
3 Study cases of MSA > 10% and < 30%, and analyze their individual consideration of other factors.
4 Investigate the relationship between $C_p$ and $C_{pk}$ when the measurement average is off from the specification center by less than one standard deviation.
5 "One argues that $P_p$ and $P_{pk}$ are not appropriate for prototyping and small tests." Please review this statement, and justify why you agree or disagree.

**6** Develop a quality and readiness checklist for a production tryout.

**7** Search a research paper on improving change management from library databases, or https://scholar.google.com, and discuss the methods used in the paper.

**8** Study key factors in making changes early and reducing the number of changes during development.

**9** Study the cost-effectiveness of an internal audit preparing for a formal third-party audit.

**10** Find a case to discuss if a quality system audit is value-added to ultimate customers.

# References

AHRQ. (n.d.) Understanding quality measurement, Agency for Healthcare Research and Quality, https://www.ahrq.gov/professionals/quality-patient-safety/quality-resources/tools/chtoolbx/understand/index.html, accessed in May 2020.

AIAG. (2006). *Production Part Approval Process, PPAP*, 4th edtition. Southfield, MI: Automotive Industry Action Group.

AIAG. (2010). *Measurement System Analysis Reference Manual*, 4th edition. Southfield, MI: Automotive Industry Action Group.

Amiruddin, A. (2018). Internal quality audit in the implementation of quality assurance system of continuing education at junior high school, *Proceedings of the 2nd International Conference on Education Innovation*, Universitas Negeri Surabaya, Indonesia, July 28, 2018. 10.2991/icei-18.2018.127.

ASQ. n.d. Quality glossary, American Society for Quality, https://asq.org/quality-resources/quality-glossary, accessed in May 2020.

ASTM. (2017). *E2782-17, Standard Guide for Measurement System Analysis (MSA)*. West Conshohocken, PA: ASTM International.

Comuzzi, M. and Parhizkar, M. (2017). A methodology for enterprise systems post-implementation change management, *Industrial Management & Data Systems*, Vol. 117, Iss. 10, pp. 2241–2262. 10.1108/IMDS-11-2016-0506.

CFR. (2020). §820.22 Quality audit, PART 820—Quality system regulation, Title 21, *The Code of Federal Regulations (CFR)*, https://www.ecfr.gov/cgi-bin/text-idx?SID=c3dd10f22cb03368b8f32848d3f7e8e7&mc=true&node=pt21.8.820&rgn=div5#se21.8.820_122, accessed in July 2020.

Donatella, P., Haraldsson, M., and Tagesson, T. (2018). Do audit firm and audit costs/fees influence earnings management in Swedish municipalities, *International Review of Administrative Sciences*, Vol. 85, Iss. 4, pp. 673–691. 10.1177/0020852317748730.

Erdmann, T.P., Does, R., and Bisgaard, S. (2010). Quality quandaries: A gage R&R study in a hospital, *Quality Engineering*, Vol. 22, pp. 46–53. 10.1080/08982110903412924.

Fastowicz, J., Lech, P., and Okarma, K. (2020). Combined metrics for quality assessment of 3D printed surfaces for aesthetic purposes: Towards higher accordance with subjective evaluations, *ICCS 2020 Lecture Notes in Computer Science*, Vol. 12143, 10.1007/978-3-030-50436-6_24.

FDA. (2009). Q10 pharmaceutical quality system - Guidance for industry, U.S Food and Drug Administration, https://www.fda.gov/regulatory-information/search-fda-guidance-documents/q10-pharmaceutical-quality-system, Content current as of 10/17/2019, accessed in September 2020.

FDA. (2014a). Management controls, US Food and Drug Administration, https://www.fda.gov/inspections-compliance-enforcement-and-criminal-investigations/inspection-guides/management-controls, Content current as of 09/08/2014, accessed in July 2020.

FDA. (2014b). Sampling Plans, US Food and Drug Administration, https://www.fda.gov/sampling-plans, Content current as of 09/ 04/2014, accessed in January 2021.

FDA. (2019). Questions and answers on current good manufacturing practices – Production and process controls, US Food and Drug Administration, https://www.fda.gov/drugs/guidances-drugs/questions-and-answers-current-good-manufacturing-practices-production-and-process-controls#5, Content current as of 07/12/2019, accessed in September 2020.

Guliyev, V., Hajiyev, N., and Guliyev, F. (2019). Factors affecting audit quality in corporate sector, *37th International Scientific Conference on Economic and Social Development – Socio Economic Problems of Sustainable Development*, Baku, February 14–15, 2019.

Hamraz, B., Caldwell, N. and Clarkson, J. (2013). A Holistic categorization framework for literature on engineering change management, systems engineering, Wiley Online Library. 10.1002/sys.21244.

ISO. (2015a). *ISO 10012:2003 Measurement Management Systems – Requirements for Measurement Processes and Measuring Equipment, Reviewed and Confirmed in 2015*. Geneva: International Organization for Standardization.

ISO. (2015b). *ISO 9000: 2015 Quality Management Systems – Fundamentals and Vocabulary*. Geneva: International Organization for Standardization.

ISO. (2018). *ISO 19011:2018 Guidelines for Auditing Management Systems*. Geneva: International Organization for Standardization.

ISO. (2020). *ISO 9001: 2020 ISO 9001 Auditing practices group guidance on: REMOTE AUDITS*, Edition 1, Date: 2020-04-16, International Organization for Standardization and International Accreditation Forum. https://committee.iso.org/files/live/sites/tc176/files/documents/ISO%209001%20Auditing%20

Practices%20Group%20docs/Auditing%20General/APG-Remote_Audits.pdf, accessed in February 2021.

Jayatilleke, S. and Lai, R. (2018) A systematic review of requirements change management, *Information and Software Technology*, Vol. 93, pp. 163–185. 10.1016/j.infsof.2017.09.004.

Knackstedt, S.A. (2017). A case study on part engineering change management from a development and production perspective at a major automotive OEM, All Theses. 2646. https://tigerprints.clemson.edu/all_theses/2646, accessed in April 2020.

Kotz, S. and Lovelace, C., eds. (1998). *Process Capability Indices in Theory and Practice*. London: Hodder Education Publishers.

Lahidji, B. and Tucker, W. (2016). Continuous quality improvement as a central tenet of TQM: History and current status, *Quality Innovation Prosperity*, Vol. 20, Iss. 2, pp. 157–168. 10.12776/qip.v20i2.748.

Lakymenko, N., Romsdal, A., Alfnes, E. et al. (2020). Status of engineering change management in the engineer-to-order production environment: Insights from a multiple case study, *International Journal of Production Research*, Vol. 58, Iss. 15, pp. 4506–4528. 10.1080/00207543.2020.1759836.

Lellis, A.D., Leva, A., and Sulisa, E. (2018). Simulation for change management: An industrial application, *Procedia Computer Science*, Vol. 138, pp. 533–540. 10.1016/j.procs.2018.10.073.

Magalhães, S., Arieira, C., Carvalho, P. et al. (2020). Colon Capsule CLEansing Assessment and Report (CC-CLEAR): A new approach for evaluation of the quality of bowel preparation in capsule colonoscopy, *Gastrointestinal Endoscopy*, 10.1016/j.gie.2020.05.062.

McIntosh, J., Alonso, A., MacLure, K. et al. (2018). A case study of polypharmacy management in nine European countries: Implications for change management and implementation, *PLoS ONE*, Vol. 13, Iss. 4, p. e0195232, 10.1371/journal.pone.0195232.

Montgomery, D.G. (2013). *Introduction to Statistical Quality Control*. Hoboken, NJ: Wiley.

Murdock, H. (2019). *Auditor Essentials: 100 Concepts, Tools, and Techniques for Success*. Boca Raton, FL: CRC Press.

Murphy, S.A., Moeller, S.E., Page, J.R. et al. (2009). Leveraging measurement system analysis (MSA) to improve library assessment: The attribute gage R&R, *College & Research Libraries*, Vol. 70, Iss. 6, pp. 568–577.

Pan, J., Kuo, T., and Bretholt, A. (2010). Developing a new key performance index for measuring service quality, *Industrial Management & Data Systems*, Vol. 110, Iss. 6, pp. 823–840. 10.1108/02635571011055072.

Puga-Leal, R. and Pereira, A.L. (2007). Process capability in services, *International Journal of Quality & Reliability Management*, Vol. 24, Iss. 8, pp. 800–812. 10.1108/02656710710817090.

Refaat, R. and El-Henawy, I.M. (2019). Innovative method to evaluate quality management system audit results' using single value neutroso.phic number, *Cognitive Systems Research*, Vol. 57, pp. 197–206. 10.1016/j.cogsys.2018.10.014.

Reid, L.C., Carcello, J.V., Li, C. et al. (2019). Impact of auditor report changes on financial reporting quality and audit costs: Evidence from the United Kingdom, *Contemporary Accounting Research*, Vol. 36, pp. 1501–1539. 10.1111/1911-3846.12486.

Rich, F.A. and Kluse, C. (2018). Case study – Keeping things running, *ASQ Six Sigma Forum Magazine*, Vol. 17, Iss. 4, pp. 15–21.

Salehi, M., Fakhri Mahmoudi, M.R., and Daemi Gah, A. (2019). A meta-analysis approach for determinants of effective factors on audit quality: Evidence from emerging market, *Journal of Accounting in Emerging Economies*, Vol. 9, Iss. 2, pp. 287–312. 10.1108/JAEE-03-2018-0025.

Schuh, G., Gartzen, T., Soucy-Bouchard, S. et al. (2017). Enabling agility in product development through an adaptive engineering change management, *Procedia CIRP*, Vol. 63, pp. 342–347. 10.1016/j.procir.2017.03.106.

Storbjerg, S.H., Brunoe, T.D., and Nielsen, K. (2015). Towards an engineering change management maturity grid, *Journal of Engineering Design*, Vol. 27, Iss. 4–6, pp. 361–389. 10.1080/09544828.2016.1150967.

Turunen, R.M. and Watson, G.H. (2021). Modern approach, *Quality Progress*, Vol. 54, Iss. 2, pp.14–21.

Walfish, S. (2007). Statistically rational sampling plans for audits, the auditor, December 2007 e-Edition, 4 pages. http://statisticaloutsourcingservices.com/Auditor.pdf, accessed in January 2021.

Wilberg, J. (2015). Using a systemic perspective to support engineering change management, *Procedia Computer Science*, Vol. 61, pp. 287–292. 10.1016/j.procs.2015.09.217.

Zheng, P., Chen, C., and Shang, S. (2019). Towards an automatic engineering change management in smart product-service systems–A dsm-based learning approach, *Advanced Engineering Informatics*, Vol. 39, pp. 203–213. 10.1016/j.aei.2019.01.002.

# 8

# Quality Management Tools

Many tools, methods, and techniques can be used in quality management. Most of them are general, and can be used for almost all types of quality work, while some of them may be more applicable to just one of the three areas (quality planning, quality control, and continuous improvement) of a quality management system (QMS). In this chapter, from a problem-solving standpoint, the commonly used quality tools are discussed based on their unique principles and applications.

## 8.1 Problem-solving Process

Problem solving is a process, comprising a series of actions, including defining a problem, analyzing its root causes, finding and implementing a solution, and verifying the result. During the development and operations of a product, service, or process, various types of problems and issues can occur. Following a structured process and using appropriate tools are key to effective problem solving.

### 8.1.1 Plan–Do–Check–Act Approach

#### Principle of Continuous Improvement

Routine quality work can be seen as a problem-solving process. Conventional problem solving is normally reactive in nature. A traditional mindset is "if it ain't broke, don't fix it" (Martin n.d.), meaning if something is working well, leave it alone. However, such a mindset may not work anymore in today's competitive, rapidly evolving business environment. A concordant mindset and practice is continuous improvement (CI), which is similar to "if it ain't broke, *make it better*." That is, never be satisfied with the status quo, which is a necessary condition to survive in today's competitive world.

*Quality Planning and Assurance: Principles, Approaches, and Methods for Product and Service Development*, First Edition. Herman Tang.
© 2022 John Wiley & Sons, Inc. Published 2022 by John Wiley & Sons, Inc.

One widespread name for CI is *Kaizen (改善)*, which is the Sino-Japanese word for "changing for better." Kaizen was first practiced in Japanese business after World War II. Now, it is a popular word in quality fields worldwide, for the process and techniques of CI and problem solving.

It is important to note that regardless of its reactive or proactive nature, problem solving and CI follow the same general process, with some variation when using different methods and tools. The overall flow and major steps are shown in Figure 8.1. In the cycle of problem solving and CI, the third step is to find primary root causes and influencing factors, respectively. Although the name of this step differs, its process is the same for the identification of root causes and influencing factors.

A fundamental difference between ordinary problem solving and CI is mentality. For an organization, it can be about corporate culture. Fighting the inertia and reluctance of the "if it ain't broke, don't fix it" mindset can be more important than implementing technical methods and tools in many cases. Thus, Kaizen or CI can be characterized as a culture, process, and project.

From a perspective of project management, problem-solving activities are usually reactive in nature. Thus, problem-solving projects are typically assignments involving problems that have already happened. For CI purposes, one can often choose what targets to improve. Figure 8.2 shows a two-dimensional evaluation chart for the evaluation of multiple CI proposals. CI projects may be evaluated based on their sizes, benefits, resources required, technical difficulty, etc. It is recommended that new practitioners consider simple projects with visible benefits, as a starting point, to build skills and credentials, and then take on greater challenges to resolve large and complex issues.

Performance or achievement of a problem-solving and CI project may or may not have an immediate financial implication. One should try to align a problem-solving and CI project to business objectives, e.g. customer satisfaction, on-time

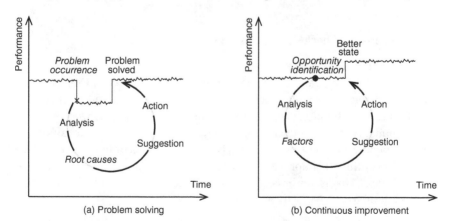

(a) Problem solving  (b) Continuous improvement

**Figure 8.1** Problem-solving and CI processes.

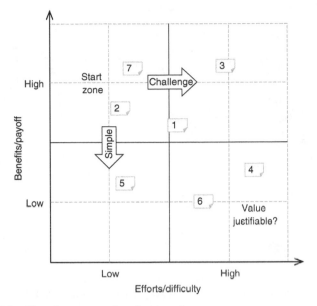

**Figure 8.2** CI project proposal evaluation chart.

delivery, market share improvement, etc. Sometimes, improving work efficiency and effectiveness may be converted into cost avoidance for a long run. Therefore, immediate financial benefits should not be an objective for every problem-solving and CI project.

### Process of Plan–Do–Check–Act

Plan–do–check–act (PDCA) is a widely used process for problem solving and CI (see Figure 8.3). PDCA is also called a quality circle. For a specific problem or improvement target, the four sequential steps in a loop are:

1. <u>Plan</u>: To recognize an improvement opportunity or identify a problem, establish an aim, and determine the methods to use. Other elements of a project, including the resources, timing, and target criteria, need to be decided as well.
   - The aim (or target) should be specific, measurable, achievable, relevant, and time bound (SMART) goal setting for a PDCA project.
   - The methods and tools selected may set the direction of the following steps and affect the likelihood of success.
   - Depending on the size and scope of a project, planning may take considerable time and effort.

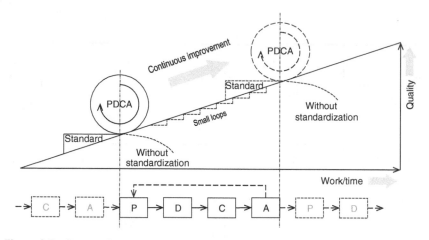

**Figure 8.3** Process flow and functionality of PDCA.

2. <u>Do</u>: To take an action, for example, making physical changes, or executing a test, based on the plan. Sometimes a smaller trial implementation can be helpful before doing full-scale planned work. For a large and complex project, smaller loops or incremental efforts can be helpful for staying on the track toward the ultimate goals, as learning is part of practice.

3. <u>Check</u> (or study): To collect data from the tasks executed, analyze the data, and compare the results between the previous status and the results after the "do" actions. If the results are not satisfactory, then the project team may need to go back to plan and/or do phases to revise. This check step, with the previous do step, needs certain analysis tools (in the following sections) to complete the work.

4. <u>Act</u> (or adjust): To identify the root causes, document lessons learned, and implement the solutions, based on the study and check results. If the project team achieves its original goals, then the next step is to set a new baseline, and update respective standards as applicable. After completing this PDCA cycle, then the project team plans the next cycle of PDCA from this new baseline. As an iterative process, a PDCA project continues until all results are genuinely satisfactory.

PDCA is a simple and effective approach, and can be used for most types of problem-solving and CI projects. PDCA is often used for small projects, and as the incremental improvement process for a larger project. A large PDCA may break into a few smaller PDCAs' to be more efficient and less demanding. However, as a structured problem-solving process, PDCA itself does not guarantee success. A successful PDCA project depends on team commitment, resources, tools used, management support, etc.

## Application of PDCA

To illustrate the four steps and their pertinence to different situations, five examples are summarized in Table 8.1. From these examples, one can see that the PDCA process can be used for different purposes, and combined with other problem-solving tools. Readers can search for specific applications in more related areas.

## Process of Define–Measure–Analyze–Improve–Control

Similar to PDCA, there is another popular problem-solving and CI process called define–measure–analyze–improve–control (DMAIC). DMAIC is also an iterative process with five steps, as shown in Figure 8.4.

The first step "define" is functionally similar to the "plan" in PDCA. The defined objectives should follow the SMART goal setting mentioned earlier. In this stage, the boundaries of the problem should also be defined, which is tied closely to the problem statement. Along with the definition of the problem to solve, one also considers the customer expectations and benefits during this step.

The second step "measure" is to collect data for the problem identified in the first step, including the status data, defects, and/or performance information. The data type, sample size, and measurement system (discussed in Chapter 7) should be addressed as well in this step. With this step, a DMAIC process relies more on being data driven than a PDCA process.

The third "analyze" step is to identify root causes, by analyzing inputs, control factors, noise factors, and improvement opportunities. Various (often statistical) data analysis methods can be used in this step. Hopefully, the root causes can be found through data analysis. If there are multiple root causes, they should be ranked, and the corresponding analysis effort should be prioritized. For variation reduction projects, the analysis is focused on the assignable sources of variation.

The following step is to "improve" the situation by fixing the root causes and preventing the problems from reoccurring. The improved status should be reevaluated. Most times, improve actions require innovative solutions and additional resources. Temporary solutions are not recommended for many cases.

The last step is to "control" the improvement made to keep the process on its new course, by implementing the validated solution. The actual control approaches to use are case dependent, such as how preventive mistake-proofing (sometimes called *poka-yoke*) can be implemented for manual operations. At this step, a summary document with lessons learned and new ideas is compiled, for the next cycle of a DMAIC process.

DMAIC represents an overall process flow, and its five steps may not always be fully sequential. Sometimes, having a loop between the measure and the define phases may help clarify, and/or revise, the defined objectives of a project. For example, during a project, one may find that some data are not available,

**Table 8.1** Examples of PDCA applications

| Example | Plan | Do | Check | Act |
|---|---|---|---|---|
| Quality inspection: acceptance of raw materials for design of flat shoes quality (Sari et al. 2019) | To prepare quality guidelines, make check sheets for the materials, and inspection instructions | To complete the check sheets according to the inspection instructions | To check the filled check sheets and analyze the results | To separate the defect materials to return or repurchase to suppliers |
| Quality preparation: a surgical safety checklist (Alpendre et al. 2017) | To disseminate research project, request authorization, and define activities | To prepare a checklist for the pre- and post-operative with applicability to nursing care practice | To apply the checklist to verify content and to perform improvements to the checklist after the test | To validate the checklist by the Committee to another research |
| Problem solving: loss of sauce in a food process (Júnior and Broday 2019) | To solve the problem and collect process data | To carry out the proposed actions: tests of materials and equipment | To verify the current situation using the Ishikawa diagram, to figure out and confirm root causes | To apply the effective actions/solutions to production lines |
| Quality evaluation: assessment and acquisition of surgical knowledge (Jin et al. 2012) | Supervisor: to evaluate the knowledge of trainee | To perform literature search and literature "work up" | Supervisor: to identify lack of knowledge<br>Trainee: To compare own surgical images and analysis | To proceed to training if not meeting the predefined criteria |
| Development: clinical formulation (Hanawa and Momo 2019) | To identify the problems facing the field grasping points | To preparing clinical products and use | To evaluate treatment effects | To have a concept for developing new clinical products |

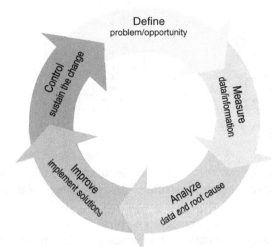

**Figure 8.4**   DMAIC process flow.

which can change the initial defined objects. Measurement tasks might also go into later phases, to play the roles of support and verification.

Because a DMAIC process emphasizes CI and intensive data analysis, it has been widely used in CI projects. As DMAIC and PDCA are similar, one may use either of them for problem solving and CI. There are other similar processes; see define–measure–analyze–design–verify (DMADV), for example, for product design. The next section has more detail on these approaches.

### 8.1.2   8D Approach

#### Process of 8D

An 8D approach is named for its eight disciplines, or eight steps of problem solving (see Figure 8.5). From a systems viewpoint, an 8D process is similar to the PDCA and DMAIC discussed earlier; however, an 8D process has more steps and functions than either of those processes. Table 8.2 explains the functions of the eight disciplines.

#### Discussion of the 8D Process

An advantage of this 8D approach is that it considers both temporary fixes and permanent corrections. The temporary fix step (D3) can be especially helpful to avoid severe consequences, e.g. costly downtime, when a permanent solution is

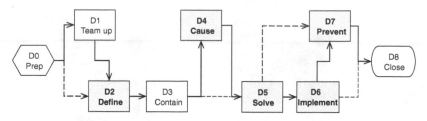

**Figure 8.5** 8D/5D problem-solving process flow.

not immediately available or known (for complex operations like vehicle manufacturing). A temporary fix buys some time to find root causes, plan, and implement a permanent solution.

Some of the eight steps are influenced by others. For example, a temporary containment at D3 may be an option for a permanent solution at D5. Similarly,

**Table 8.2** Tasks in the 8D problem-solving process.

| Discipline | Task and Explanation |
| --- | --- |
| D0 | To determine the necessity of an 8D process and plan for problem solving. |
| D1 | To form a project team of people with responsibilities and experience, to allocate time and authority, and to have the commitment of management support. |
| **D2** | **To define a problem, to state it concisely and describe it specifically, such as who, what, where, when, why, and how, for the problem.** |
| D3 | To find and implement an interim plan to contain the problem, to isolate the problem from customers, and to validate the improved results from the interim action. |
| **D4** | **To identify the root causes of the problem and verify them. The team may use various tools, e.g. a fishbone diagram and statistical analysis.** |
| **D5** | **To develop a permanent corrective solution for the problem. The solution must remove the root causes identified. If multiple corrective proposals are available, to take other factors, e.g. effect, risk, cost, and time, into account for the proposal selection.** |
| **D6** | **To implement the corrective plan and verify its results, and remove the interim action in D3 if it is different. To monitor the effect of the permanent solution for a little while.** |
| **D7** | **To develop measures to prevent the recurrence of this problem and similar ones, which may include the modification of a system, procedure, and/or work instruction.** |
| D8 | To close the problem-solving process and recognize the team. |

a permanent solution at D5 can be a preventive solution at D7. Because of the complexity of a problem and its root causes, preventive solutions may or may not be always available. In other words, 100% prevention from recurrence sometimes may be an impractical expectation.

Step D4, root cause analysis, plays the central part of the 8D process; because there is no hope for an 8D success without knowing the problem's root causes. This step often requires a lot of analysis. Various technical methods, including the 14 tools discussed in the following sections, may be used.

The 8D approach was originally used in automotive manufacturing, but has now been applied in other industries like product design (Fritsche 2016), healthcare (Faloudah et al. 2015), and others (Aras and Özcan 2016). Most 8D projects are internal, and remain unpublished.

8D problem solving is a well-structured process, but can be time consuming to complete. From time to time, some steps are unnecessary for small and simple problems, so a simplified version of 8D is used. This is called 5D, which has and contains the five steps highlighted in Figure 8.5 and Table 8.2. The D2 define, D4 cause, D5 solve, D6 implement, and D7 prevent can be considered the core elements of the 8D, which are also comparable to the PDCA and DMAIC processes discussed.

### A3 and Other Approaches

Another problem-solving process is called A3, because the information of an entire problem-solving project is documented on an A3-sized piece of paper (297 mm × 420 mm or 11.7 in × 16.5 in). In addition, the A3 process is very similar to 8D in terms of the overall process, detail steps, and documentation. The A3 process was first used by Toyota (Sobek and Jimmerson 2004) and has since been used in many industries. A typical layout of the A3 form is shown in Figure 8.6.

Characteristic of the A3 approach is its one-page documentation for most problem-solving activities, which is often preferred, as this format summarizes the key information of problem solving into one page for effective communication. It is a unique characteristic that a one-page A3 is often used at the point of the issue on the shop floor to get workers involved. This way, everyone can "go and see" when working on a problem using an A3 form.

In an A3 report, brief pictures, diagrams, and charts may be included if they fit onto the one pager. The associated detail data and evidence can be attachments. For example, one report provided the detailed implementation of how an A3 approach was used to improve the $CO_2$ emission of a company transportation (Lenort et al. 2017).

Many other approaches can be used in problem-solving and CI processes. Some of these approaches are simpler, while others are more complicated. For example, a similar process to PDCA is called DIVE (define–investigate–verify–ensure). The overall processes of PDCA and DIVE are about the same, while the individual steps of DIVE have different scopes of work and focuses.

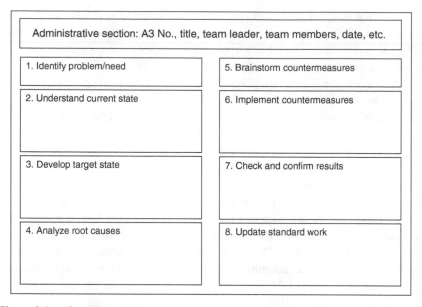

**Figure 8.6** A3 problem-solving process template.

Advanced problem-solving approaches and processes use unique technologies and have special focuses. For example, the Kepner-Tregoe method comprises four types of analysis on: 1) problem situation, 2) causes and their interrelations, 3) solution decision, and 4) future prevention. Another method is called TRIZ (translated from Russian as the theory of inventive problem solving). The TRIZ process emphasizes the analysis of technical systems for understanding and solving complex management problems. The TRIZ is an 85-step procedure to solve complicated invention problems, and TRIZ-dedicated software is available. Note that using the advanced problem-solving methods often requires specific training.

### Abstract of Problem-solving Processes

Reviewing common problem-solving approaches, e.g. PDCA, DMAIC, 8D, and A3, one can recognize that they have similar process flows, with moderately different levels of both detail and overlaps (Figure 8.7).

In a problem-solving and CI process, regardless of which approach is used, the three core elements are problem identification, root cause analysis, and solution implementation. The basic workflow can be summarized into three overlapped stages, with some unique characteristics for a specific approach:

1. Problem understanding and identification. The tasks in this stage normally include descriptively and qualitatively describing a problem, and doing plan-

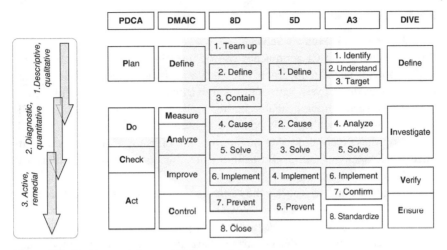

**Figure 8.7** Problem-solving and CI approaches.

ning work. An identified problem may be phrased in different ways, which reflects the work focus or method to use.

2. <u>Root cause analysis</u>. This stage is diagnostic in nature. A quantitative analysis is strongly preferred to find the root causes, because it is more objective and accurate in most cases (Tang 2021). Sometimes, an analysis stays qualitative because of data availability.

3. <u>Solution implementation</u>. In this stage, a solution is proposed based on the root cause analysis. Solution implementation (actions) and result verification are the key functions in this stage.

One method can be more popular than another in a given field, but not necessarily more superior. Because of their similarity, selecting a problem-solving process remains a personal judgment call. The success of problem solving relies on various factors, such as data reliability, team efforts, technical capability, and available resources. In addition, a problem-solving process ties into many other methods and tools. In other words, a problem-solving process and tools must be well integrated to solve problems efficiently.

Mentioned in the discussion of A3, any project document of problem solving and CI should be displayed and maintained at the area where a problem is addressed. An on-site display is an effective way to encourage the participation of frontline personnel.

Sometimes a lot of effort is put into problem-solving documentation, just for an impressive appearance. This kind of practice may be useful for training and reporting purposes, but does not necessarily add value to the process effectiveness itself, or its achievements. Making problem-solving documentation effective (not just impressive) remains a concern to many practitioners.

### 8.1.3 Approaches and Tools

#### Tools Applications in Problem Solving

Problem-solving and CI approaches can be used in conjunction with several tools. Applying such an approach, one must also use appropriate tools to identify problems, do analysis, figure out the root causes, and generate solutions.

There are various tools available in the "toolbox" of problem solving and CI. Table 8.3 lists 14 quality tools commonly used in problem solving and CI. Mastering these tools and using the appropriate ones for a situation can optimize problem-solving and CI efforts. In the next two sections, these 14 tools will be reviewed one by one.

There can be many caveats for the common uses listed in the table. For example, a Pareto chart (#4 in the table) may be used in the "plan" phase of a project, as long as there is appropriate data. For example, with defect data or warranty data, one can generate a defect Pareto chart and plan tests for major issues accordingly.

If an analysis tool is needed in a process planning, the tool may be more applicable to certain types of development tasks, too. As a general reference, Table 8.4 shows the main suitable applications of these tools in the development phases of a product or service.

**Table 8.3** Common applications of quality tools in PDCA process

| Tool | Plan | Do | Check | Act |
|------|------|-----|-------|-----|
| 1. Fishbone diagram | √ | √ | | |
| 2. Check sheet | √ | √ | √ | √ |
| 3. Histogram | | √ | √ | |
| 4. Pareto chart | | √ | √ | |
| 5. Scatter diagram | | √ | √ | |
| 6. Control charts | | √ | √ | √ |
| 7. Stratification | | √ | √ | |
| 8. Affinity diagram | | √ | √ | |
| 9. Relations diagram | √ | √ | | |
| 10. Tree diagram | | √ | √ | |
| 11. Matrix chart (diagram) | √ | √ | | |
| 12. Network diagram | √ | √ | √ | |
| 13. Prioritization matrix | √ | √ | √ | √ |
| 14. Process decision program chart | √ | √ | | |

**Table 8.4** Applications of quality tools in product/service development

| Tool | Market | Concept | Design | Plan | Test | Launch |
|---|---|---|---|---|---|---|
| 1. Fishbone diagram | √ | √ | | | √ | √ |
| 2. Check sheet | | | √ | √ | √ | √ |
| 3. Histogram | √ | | | | √ | √ |
| 4. Pareto chart | √ | √ | | √ | √ | √ |
| 5. Scatter diagram | | | | | √ | √ |
| 6. Control charts | | | | | √ | √ |
| 7. Stratification | √ | | √ | | √ | |
| 8. Affinity diagram | √ | √ | | | √ | √ |
| 9. Relations diagram | √ | √ | √ | √ | | |
| 10. Tree diagram | | √ | √ | √ | | |
| 11. Matrix chart (diagram) | | √ | √ | √ | | |
| 12. Network diagram | | √ | | √ | √ | |
| 13. Prioritization matrix | √ | | √ | √ | √ | √ |
| 14. Process decision program chart | √ | √ | √ | √ | | |

**Additional Tools**

In addition to these 14 tools, many other tools can be used for problem-solving processes. For example, one simple tool is called 5W1H, as the abbreviation of six questions for understanding of a problem. 5W1H can be used in a form format (Table 8.5), where the "is not" answer is optional to describe what the problem could be reasonably but is not. 5W1H can be effective for an initial problem identification and preliminary root cause analysis.

Similarly, another common tool is called "5-whys": that is to ask why for the nature and reason of a problem multiple times. Most frontline people like 5-whys as a simple, effective tool for root cause identification at the beginning. Note the replies to a why question should be factual, backed by data, and specific. The problem-solving team should then determine the next why to ask. There is no need to ask why exactly five times.

Another tool for problem identification is SIPOC (supplier, inputs, processes, outputs, and customers) discussed in Chapter 1. For an issue, a team may discuss and identify these five elements of SIPOC, which helps qualify the entire process and find the key factors associated with a given issue.

There are many other tools in the toolbox of problem solving and CI. Briefly reviewed in Tables 8.3 and 8.4, each tool has its characteristics and purposes,

**Table 8.5** 5W1H form for problem identification and selection.

| Category | Typical question | Answer (is) | Answer (is not) |
|---|---|---|---|
| What | What problem does a process, product, or system have? <br> What is exactly the problem? | | |
| Where | Where is the problem located geographically? <br> Where is the problem within the object? | | |
| When | When was the problem first detected? <br> When, since the first time, has the problem been observed again? | | |
| Who | Who detected the problem? | | |
| Why | Why did the problem happen at this particular time/location? | | |
| How or How many | How was the problem detected? <br> How many defects are in a process, product, or system | | |

while a task can have multiple tools applicable. Therefore, selecting proper tools for a specific task is a key to the effectiveness of problem solving and CI.

**Qualitative, Quantitative, and Other Analyses**

As discussed in Chapter 1, the quality management is data driven. Data may be categorized into two types: qualitative (attribute) and quantitative (variable) data. A particular method is often primarily suitable for one type of data. For example, a fishbone diagram is a brainstorming and qualitative tool that is widely used for analyzing unknown factors in the early phases of a problem-solving process. Sometimes, one would need an appropriate tool to find relationships and prioritize tasks. Table 8.6 is another reference for the 14 tools for root cause analysis in problem solving and CI.

A qualitative analysis is often used to describe a problem or root cause, but it is not able to quantify it in most cases. Thus, a qualitative analysis is normally conducted in the early phases of problem solving. A quantitative analysis, requiring numerical data, is more objective, and can go further toward root cause analysis. For a large or complex problem, a quantitative analysis is highly recommended if corresponding data is available.

In a problem-solving project, a process and quality tools are normally integrated. Quality professionals should be knowledgeable about this integration. Other factors, such as project objectives and people's skills, should be considered when selecting methods, shown in Figure 8.8 (Tang 2021).

**Table 8.6** Typical analysis of using quality tools

| Tool | Qualitative analysis | Quantitative analysis | Relationship analysis | Prioritization analysis |
|---|---|---|---|---|
| 1. Fishbone diagram | √ | | √ | |
| 2. Check sheet | √ | √ | | |
| 3. Histogram | | √ | | √ |
| 4. Pareto chart | | √ | | √ |
| 5. Scatter diagram | | √ | √ | |
| 6. Control charts | √ | √ | | √ |
| 7. Stratification | | √ | √ | √ |
| 8. Affinity diagram | √ | | | √ |
| 9. Relations diagram | √ | | √ | |
| 10. Tree diagram | √ | √ | √ | |
| 11. Matrix chart (diagram) | √ | | √ | |
| 12. Network diagram | | √ | √ | √ |
| 13. Prioritization matrix | | √ | √ | √ |
| 14. Process decision program chart | √ | | √ | √ |

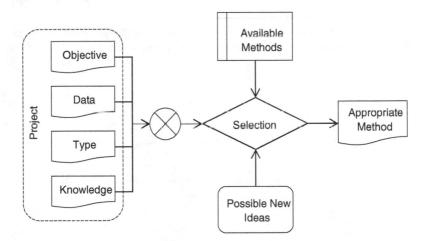

**Figure 8.8** Considerations in problem-solving method selection. *Source*: Tang, H., (2021). *Engineering Research – Design, Methods, and Publication*, ISBN: 978-1119624486, Hoboken, NJ: Wiley.

Frequently, a specific process and particular tool is predetermined as a common practice, or required by a business procedure. While a predetermined, proven tool is suitable, it may or may not be the best effective method for a specific problem. Being open minded to adopting different or even creating new methods is essential for quality professionals to execute their duties productively.

## 8.2 Seven Basic Tools

The seven basic quality management tools are a fishbone diagram, check sheet, histogram, Pareto chart, scatter diagram, control charts, and stratification. They have been proven effective, and widely used in various industries.

### 8.2.1 Cause-and-effect Diagram

A cause-and-effect diagram, also called a fishbone diagram because of its similarity to a fish's skeleton, is often used to find factors for a problem based on cross-functional team brainstorming activities. In general, six types of factors or causes contribute to a problem (or effect), shown in Figure 8.9. For a specific problem, these factors do not equally contribute to the effect, or there are additional factors. For example, for a marketing issue, the factor can be product, pricing, place, promotion, people, positioning, and packaging. Thus, an investigation often tries to find the primary factor(s) based on the nature and characteristics of the problem identified.

Many case studies have used fishbone diagrams to address unique situations. For example, a fishbone diagram with five dimensions was used for a service

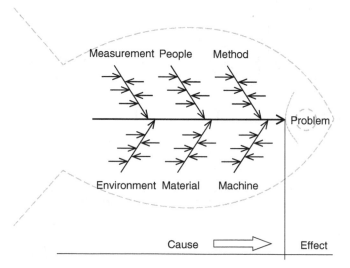

**Figure 8.9** General cause-and-effect (fishbone) diagram.

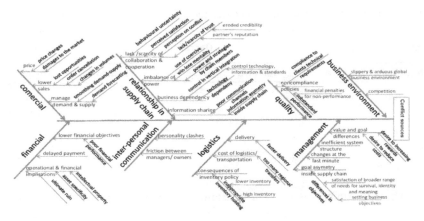

**Figure 8.10** Example of fishbone diagram application. *Source*: Constantinescu, G.C. (2017). Sources of Supply Chain Conflicts – A Fishbone Diagram Correlation, SEA: Practical Application of Science, Editorial Department, Iss. 13, pp. 191–197.

quality in Chapter 4 Figure 4.14. Figure 8.10 shows the sources of supply chain conflicts (Constantinescu 2017). In this study, the author considered eight factors/categories (e.g. business environment, quality, etc.) with 22 sub-categories, and even went to the third levels for some factors. From this fishbone diagram, one may derive influencing factors, sub-factors, and their paths, e.g. compliance to client's technical requirements → noncompliance policies → quality → conflict sources (effect).

A fishbone diagram shows an overall picture of the influencing factors to a specific problem, which is useful as a visual guide to a further investigation. However, there are some limitations, for example:

- Only talking about possibilities, but not singling out the actual root causes
- Produced from brainstorming based on opinions, may or may not be based on the evidence or data
- No interrelationship across factors/categories

Therefore, a fishbone diagram is often only used as a first step and base for further investigation and data collection, in problem-solving projects.

## 8.2.2 Check Sheet

A check sheet is a simple, structured form that is used to collect simple data on site, and make a quick judgment for a wide variety of purposes. The data in a check sheet can be either quantitative (numerical) or qualitative (attribute). Quantitative applications may have specific information, for example, to collect quality defects (see Tables 8.7) for their types and locations. Based on the check results, it is easy to identify which types of defects occur more often, and which

**Table 8.7** Example of a nonconforming check form.

| Date: | Supervisor: | | Operation: | | No. |
|---|---|---|---|---|---|
| Defect | Station A | Station B | Station C | Station D | Total |
| Type 1 | 3 | 5 | 1 | 4 | 13 |
| Type 2 | 6 | 8 | 10 | 12 | 36 |
| ... ... | | | | | |
| Total | 9 | 13 | 11 | 16 | 49 |

station has more quality issues. The type and location of a quality issue are often associated with their root causes. Therefore, a check sheet may provide quick information for quality corrective actions.

A check sheet is also often used for quick reporting. For a qualitative application, a checklist is used for describing status, i.e. yes, no, or NA. A check sheet may be for checking and tracking the status of multiple items, as a tangible evidence for problem identification. Table 8.8 shows an example of visual inspection records during production time. In Chapter 7, a preflight checklist (see Table 7.3 as an example) is used to confirm the readiness of a production trial run. Although such checklists are simple, they are essential and practical for many applications.

## 8.2.3 Histogram

Histogram charts are an analytical, approximate representation to show frequency distributions of data. In a histogram, the horizontal axis is about a series of intervals, which are often called "bins" or "buckets." The vertical axis is the frequency value of data in a bin.

Sizing a bin is a judgment call based on the range of horizontal data, and experience of the individual. It is a common practice that the number of bins is set between five and thirty. Figure 8.11 (a) and (b) show two histograms from the

**Table 8.8** Example of a visual inspection check form

| Date: | | Inspector: | | Process: | | No. | |
|---|---|---|---|---|---|---|---|
| No. | Time | Station A | Station B | Station C | Station D | Conclusion |
| 1 | 8:15am | √ | √ | √ | | OK |
| 2 | 10:05am | | | √ | √ | OK |
| 3 | ... ... | | | | | |

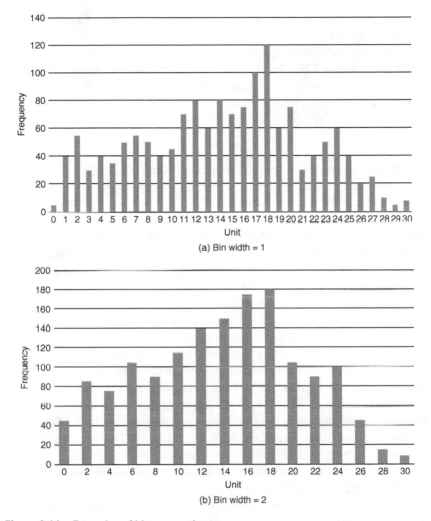

(a) Bin width = 1

(b) Bin width = 2

**Figure 8.11** Examples of histogram charts.

same data, but with different sizes of bins. To show an overall distribution, (b) may be preferred.

The histogram charts have been used in data analysis for quality assurance and guidance. For example, a histogram prediction applied to a treatment planning in healthcare (Xu et al. 2019). Some specific histograms, such as a volume histogram and distance-to-target histogram, were developed and used to evaluate quality (Roy et al. 2019). Specification limits can be set and used on a histogram chart to display the capability of a given process.

A histogram diagram is appropriate for showing the distribution of data, while ordinary bar charts can be used for other purposes, e.g. a trend of data. In addition, frequency values in histograms are relative, depending on bin size, as illustrated in Figure 8.11. The frequency values are different, and a histogram's appearance change, when bin sizes are different. Thus, it may be a good idea to try different bin sizes to see which one is most helpful to a given problem.

### 8.2.4  Pareto Chart

A problem or status of a system can be viewed as an outcome "Y" with several influencing factors, causes, or inputs called "X." Then, the relationship can be represented $Y = f(X_i)$, where f stands for function and $i = 1, 2, ....$ Because each $X_i$ may not equally affect the Y, one can figure out the relationship between Y and $X_i$ by conducting tests and data analyses. With the quantitative relationships built between $X_i$ and Y, one can generate a Pareto chart based on the amount of influence X has, as shown in Figure 8.12. The dominant X ($X_3$ in the figure) should get attention first. The Xs can be the counts of defective product, warranty cost, or customer concerns, etc.

A Pareto chart is a bar graph that shows the relative significance of factors, as their influences are represented in descending order. As a bar chart, a Pareto chart can be laid out either vertically or horizontally. The Pareto principle is sometimes referred to as the "80/20 rule." The rule suggests that 20% of causes account for 80% of the total number of quality problems, where the 80% and 20% are explanatory rather than exact numbers. Therefore, Pareto charting is a tool to sort out major elements of a complex situation for prioritization and decision. In many cases, representing data in both total number and percentage in Pareto charts can reveal important information. Subsequent resolution or improvement efforts should be dedicated to the top issues (dominant factors).

For the application of Pareto charts, one must have appropriate quantitative data. The reliability of a Pareto chart depends on the reliability of data. In addition, the results from a Pareto chart can be descriptive for a specific moment in time, but are neither prescriptive nor predictive.

The Pareto principle has been widely used across many industries. For example, a Pareto analysis was conducted for a planning model of renewable energy sources and energy storage (Li et al. 2018), and another one for multiple Pareto-optimal plans (Ceberg et al. 2018). Some similar methods, e.g. Red X, have been developed and implemented based on the Pareto principle (Schmidt 2012)). The Pareto principle itself may keep evoking smarter Pareto charts, super- and supra-Pareto charts, in a world of growing improvement of artificial intelligence and big data analytics (Schrage 2017).

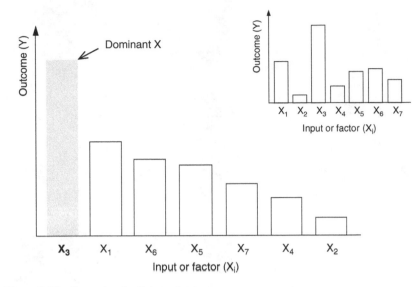

**Figure 8.12** Example of a Pareto chart.

## 8.2.5 Scatter Diagram

A scatter diagram is a two-dimension X–Y graphic for a pair of numerical data sets with one variable per axis (see Figure 8.13). Scatter diagrams are often used to identify visually whether a relationship exists between two variables, such as a variable systematically incrementing or decrementing with another. Thus, a scatter diagram relationship is often used between a cause and an effect. Occasionally, a scatter diagram is used for showing the relationship between two causes.

A scatter diagram is often used as the first step to recognize relationships, as a strong relation may often be easily identified with human eyes most times. A strong relationship suggests further analysis, such as a regression analysis and linear correlation analysis. If a relationship seems linear, then a linear regression can be used to create a mathematical expression. For linear situations, a simple, approximate index $R^2$ is often used to show how close the data fit the regression line. However, if a relationship looks nonlinear, then many other functions may be tried. MS Excel has some such functions.

Summarily, a data pattern can be represented by a regression function. The points fall along a line (linear relationship) or curve (nonlinear relationship). A linear relationship also suggests a strong correlation. Analytical software may suggest various kinds of mathematical correlations between two variables for a scatter plot. Readers can refer to dedicated books for regression and correlation analysis. Note that a mathematical correlation does not necessarily mean a causational relationship.

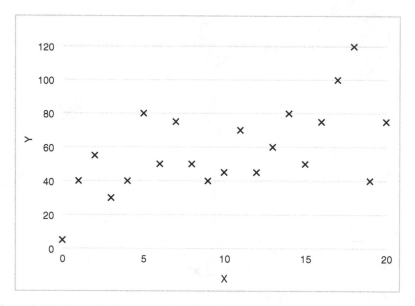

**Figure 8.13** Example of a scatter diagram.

A scatter diagram is a simple data illustration tool with no further analysis on the raw data. A scatter diagram has been commonly used, for example, on a productivity improvement (Memon et al. 2019). For three variables, 3D X–Y–Z scatter diagrams can be generated using certain software.

### 8.2.6 Control Charts

Control charts are actually a family of diagrams with various types. They are graphic illustrations to show how a process is under control or out of control based on statistics over time. Thus, control charts are often called statistical process control (SPC) charts. Many books are dedicated to SPC charts. Figure 8.14 shows examples of X-bar ($\overline{x}$) and R control charts.

Commonly used SPC charts include $\overline{x}$, R, n, c, and MA charts. As a control chart has its own specific applications, selecting appropriate control charts is important. Table 8.9 shows the characteristics and typical applications of common SPC charts as a selection reference.

Most control charts have a centerline, upper and lower control limits. For a quality feature in the modus of the lower the better, the lower limit should be zero (for example, for the number of defects per unit). If all data is within control

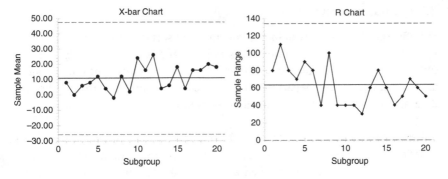

**Figure 8.14** Examples of X-bar and R control charts.

limits, then the process is deemed under control or stable. However, if it is not under control, additional analyses should be conducted to find root causes.

A control chart can also work as a monitoring tool for the state of a process over time, and gauging a necessity for further analysis and actions. Even if a process is under control, its further behavior may be predicted using certain control charts, for the consideration of possible preventive action.

**Table 8.9** Characteristics of common SPC charts

| Type of chart | Parameter | Data | Application example |
|---|---|---|---|
| $\overline{x}$ and R | $\overline{x}$ : average; R: range of data | Variable | All situations when variable data available, using R if smaller or variable sample size |
| $\overline{x}$ and s | $\overline{x}$ : average; s: sample standard deviation | Variable | |
| p | p: proportion of nonconforming units in a sample | Attribute | Defective proportion of a shipment |
| np | np: number of nonconforming units in a sample | Attribute | Number of bad results in a week |
| c | c: number (count) of nonconformities per unit | Attribute | Number of defective product returns per shift |
| u | u: average number of nonconformities per unit | Attribute | Number of new infections in a hospital per day |
| EWMA | EWMA: exponentially weighted moving average | Variable/ attribute | Detecting small shifts in a process over time |
| CUSUM | CUSUM: cumulative sum of quality characteristic measurement | Variable/ attribute | |

### 8.2.7 Stratification Analysis

Stratification is an analytical tool to separate and sort data into distinct categories, groups, or layers, based on their similarity and patterns. Quality data are often multifaceted. A typical case is that data come from different sources, such as pieces of equipment, departments, shifts, materials, and/or periods. When all data are mixed, it is difficult to identify common problems and associated causation.

Stratification analysis can be a useful method for grouping such data with descriptive and graphical representations (Figure 8.15). If data come from different sources, they can be marked and grouped with unique identifications during their collection and analysis. Stratified data can be represented in dataset lists and various graphics. Stratification analysis may lead to further data analysis.

Stratification analysis can be effective for complex and large datasets. For example, the US Centers for Disease Control and Prevention (CDC) regularly published hospitalization data during the COVID-19 pandemic. The published data are already stratified with the information of groups of people, such as ages and underlying medical conditions, as shown in Figure 8.16 (CDC 2020) as an example. Such stratified data can be a solid foundation for further studies and actions.

There are several stratification methods, including sampling, data analysis, and modeling. Stratification may be represented in data, equations, and/or figures. For example, a stratification method was used to analyze risks and predict the outcomes of patients for a specific surgery (Bateni et al. 2018).

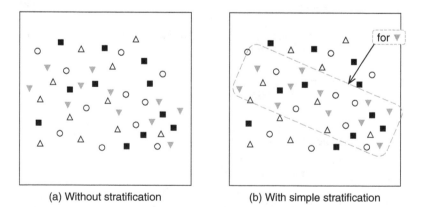

(a) Without stratification          (b) With simple stratification

**Figure 8.15**   Illustration of a stratification diagram.

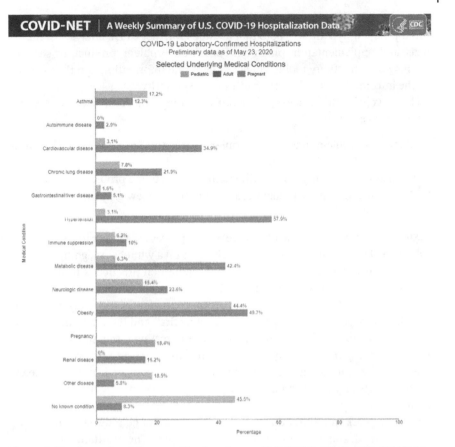

**Figure 8.16** Example of data stratification. *Source*: CDC, (2020). COVID-NET: COVID-19-Associated Hospitalization Surveillance Network, Centers for Disease Control and Prevention. https://gis.cdc.gov/grasp/COVIDNet/COVID19_5.html, accessed on June 1, 2020.

## 8.3 Seven Additional Tools

There are various other tools for quality management. Many quality practitioners use seven additional tools, in addition to the seven basic ones discussed so far. These seven additional tools are an affinity diagram, relations diagram, tree diagram, matrix chart (diagram), network diagram, prioritization matrix, and process decision program chart.

### 8.3.1 Affinity Diagram

The affinity diagram presents as a sort of "mind-mapping" technique. In a team brainstorming process, team members generate ideas about a specific problem, and then link up to other considerations to form patterns of thought. This tool can be used to organize ideas or concerns into their characteristic groups, e.g. initial problem hypotheses, potential avenues for root cause analysis, solution ideas, and implementation planning. For a new problem, product, or service, one may use an affinity tool to collect or generate information and then organize the information based on their affinity or similarity.

The development process of an affinity diagram comprises four steps (Figure 8.17):

1. Identify a problem, e.g. low customer satisfaction, to address, and form a team.
2. Brainstorm to create a list of influencing factors to the problem.
3. Discuss the affinity in all factors, and fit them into few groups.
4. Document the findings.

In Step 2, teamwork may start from creating a process of factors from individuals' inputs. Then team members post their ideas on a whiteboard, going through them to consider additional items to add. A common practice is to use sticky notes for the input, as they can be repositioned on a board for grouping easily.

In Step 3, team members discuss and group the factors, which may take some time. In teamwork, attempting to define groups first, and then determining relevance of a factor for that group may lead to interesting discussions. The team may combine, split, and reorganize the groups in a later discussion. At the end of Step 3, the team assigns the headings for the groups of factors. The team leader collects all the information, and creates a report as the affinity development output.

The teamwork in affinity diagram development can help develop important lessons learned, and benefit team building in general. The resultant affinity diagram may lead to further studies. An affinity diagram can be used for various purposes, similar to the cause–effect fishbone diagram. For example, one may use an affinity diagram during quality planning, to gather and organize thoughts and ideas. For problem-solving purposes, an affinity diagram may be used to guide a brainstorming process for the influencing factors to a particular problem. Figure 8.18 shows an example on a quality issue with yet-to-know root causes (Hughes 2015).

A similar brainstorming tool is called the nominal group technique (NGT). Unique in NGT, team members rank the importance of each item individually and silently, and then the team facilitator ranks all ideas. Thus, NGT is good for prioritizing items in decision-making.

### 8.3.2 Relation Diagram

Discussed earlier, a cause–effect fishbone diagram can illustrate multiple influencing factors and inputs. However, there is no interrelationship between the factors in a fishbone diagram. Particularly for complex problems, some factors

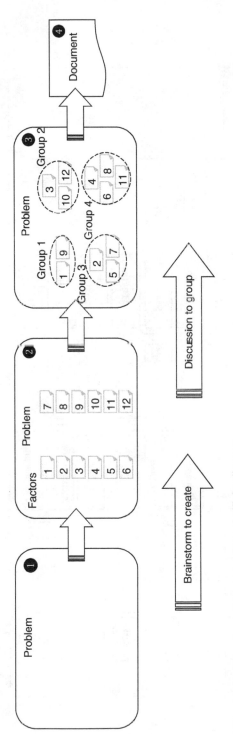

**Figure 8.17** Development process flow of affinity diagram.

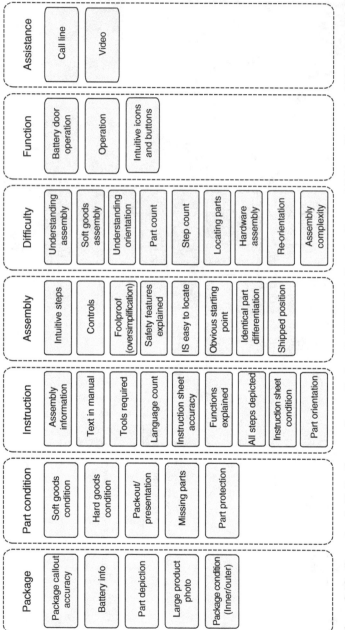

**Figure 8.18** Example of affinity diagram. *Source:* Hughes, C., (2015). Unit Exercise, Eastern Michigan University QUAL 548 Tools for Continuous Improvement (CRN 16793), November 23, 2015.

have important interrelationships, such as between the people and methods used, jointly influencing a problem. Therefore, understanding the dynamics between the factors of a problem, or between the aspects of a complex situation, is a prerequisite to solving a complex problem.

Relation diagrams are also called interrelation or network diagrams. Relation diagrams are often used to depict causal relations (see Figure 8.19). Applying a relation diagram helps analyze the natural links between aspects of a complex situation. Developing a relation diagram can be conducted in three steps:

1. Identify a problem and its factors and place them on a board, similar to the second step in affinity diagram development.
2. Connect the factors deemed having a relationship with arrow lines, which can lead to a discussion on their cause–effect relationship. Note that a factor may have multiple relationships with others.
3. Rearrange the items, with a spaghetti-like relationship, into a net diagram form.

As a relation diagram reveals the relationship between the problem and its factors, one can continue problem solving, after the factors are brainstormed in the next step. For example, Tenbergen et al. (2018) used a relation diagram to study and increase a validation objectivity of requirements-based hazard mitigations.

Unlike a fishbone diagram that can handle multiple layers of factors, a relation diagram is normally only for one layer of factors. One may take advantage

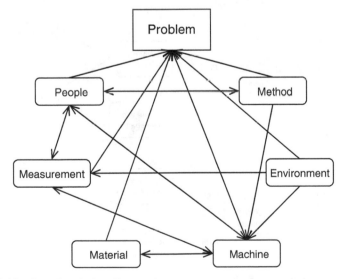

**Figure 8.19**   Generic relation diagram.

of both fishbone and relation diagrams by jointly using them. For example, a fishbone diagram can be used to show the overall causes to a problem, while using a relation diagram to illustrate the relationship among the problem and its causes.

One limitation of a relation diagram is its qualitative nature. It is possible to assign a weight factor to a relationship. However, it would need significant efforts in data analysis to support the values of weights. Otherwise, the relationship remains qualitative, and the weights assigned can be subjective.

### 8.3.3 Tree Diagram

A tree-like diagram, like a family-tree diagram or an organization chart, is to take a broad goal or idea, break it down into a few branches, and go into their finer levels. A tree diagram is in a shape of a hierarchical structure to show the connections and relationships between layers, and between siblings within a layer.

A tree diagram can be developed for goal setting, project planning, a specific problem, etc. For goal setting or project planning, one breaks it down into sub-goals or sub-plans. For a problem, one may use a tree diagram to show sub-problems or influencing factors. A tree diagram may also illustrate the elements of a product or service. A tree diagram example of business process functions is shown in Figure 1.2 in Chapter 1.

Depending on the complexity of a goal, plan, or problem, the depth of a tree diagram is often two to four layers. A tree diagram may be laid out either verti-

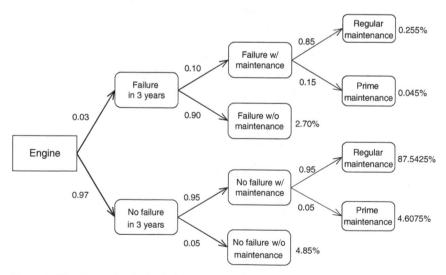

**Figure 8.20** Example of a probability tree diagram.

cally (as in top–down) or horizontally (from left to right). If an affinity diagram or a cause–effect fishbone diagram has been developed, the same information could form a tree diagram as an alternative illustration. A tree diagram can be quickly developed using MS PowerPoint or MS Visio.

A probability tree diagram, also discussed in Chapter 2, may be adapted for problem solving, which not only shows a branch structure but also includes the certain probability of each element on a given branch. Figure 8.20 shows an example of the failure probability of a new engine, where the information can be useful for the planning of maintenance and warranty repair services. Section 2.2.4 in Chapter 2 also discusses other types of tree diagrams (event tree, fault tree, and bowtie analysis) with examples.

### 8.3.4 Matrix Chart (Diagram)

A matrix chart is used to show the relationship between two or more groups, such as between objectives and methods, results and causes, tasks, or people. If there are two groups of information, then a matrix is in an L shape. When applying to three groups of information, the matrix is in a T shape. Two simple examples in L and T shapes, respectively, are shown in Figure 8.21.

The development and comprehension of a more-than-three group matrix is difficult. The quality function deployment (QFD) house of quality discussed in Chapter 3 is an example of showing this relationship among four groups of items, which is in an X shape. The relationship in a matrix chart can be quantitative (using numbers) or qualitative (using symbols). In a quantitative relation (Figure 8.21 (a)), more detailed information, e.g. the different strength and average of relations, may be illustrated in matrix diagrams.

A matrix chart can be used for different purposes, e.g. to identify the relationship between a problem and its root causes, allocate resources to a quality

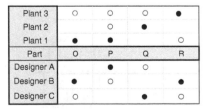

(a) L-shape, quantitative          (b) T-shape, qualitative

**Figure 8.21**   Examples of matrix diagrams.

goal, or compare potential opportunities with several variables. The functionality of a matrix chart is like other types of relation analysis tools, such as the relation diagram and tree diagram discussed earlier. Nevertheless, a matrix chart is of a different format, and is normally for simple cases. To show a relationship, one may select an appropriate one, or use them jointly, for a specific problem.

### 8.3.5  Network Diagram

A network diagram, also known as an arrow diagram, is used to represent a sequence of the activities or a required order of the tasks in a project. A network diagram is in a similar shape of a process flow diagram, and can be useful for a complex project with simultaneous tasks. To use a network diagram, one must know all requisite tasks for project completion, the duration of each task, and the predecessors of each task.

When a network diagram has multiple paths, the path that controls the entire project is called the critical path, as this route determines the project end date. A critical path is normally calculated based on the timing of all tasks by using software. Figure 8.22 depicts two styles of network diagrams with the critical path highlighted.

The network diagram can contain more detailed information, such as scheduling, prerequisites, and associated resources. Network diagraming has been widely used in project management. Microsoft has software called MS Project that can be used to create a professional looking network diagram, with a lot of information about a project. Figure 8.23 shows an example of a network diagram of new vehicle program development with the critical paths that are highlighted in red in the original graphic.

Because of the dynamics of project execution, a project network diagram should be updated periodically, e.g. weekly, so that a project manager can track and manage overall progress. Project adjustments based on a network diagram may optimize project performance and productivity.

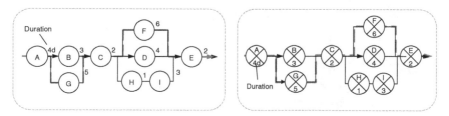

**Figure 8.22**  Illustrations of network diagrams.

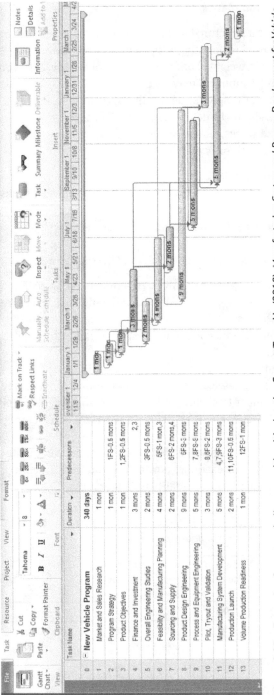

**Figure 8.23** Example of a network diagram with critical paths. *Source:* Tang, H., (2018). *Manufacturing System and Process Development for Vehicle Assembly.* Warrendale, PA: SAE International.

### 8.3.6 Prioritization Matrix

A priority matrix is a structured tool to show what factors matter most. A priority matrix can be used to quantitatively compare and prioritize items and tasks, and describe them in weighted criteria, in terms of quality, cost, benefit, timing, etc. A large array of numbers may be quickly analyzed in a matrix form. The concept and values of weight rating, similar to those in failure mode and effects analysis (FMEA), should be determined first. The outcomes of a priority matrix are these weighted scores, which may be a base for decision-making. Four steps are needed to build a priority matrix:

1. List the defined tasks.
2. Determine the factors involved, and their respective weights, based on team consensus.
3. Fill in the matrix by the rankings of each task based on a discussion, survey, or interviews.
4. Calculate the average scores and weighted average scores.

Table 8.10 illustrates a generic example of a prioritization matrix. In the table, the rating is based on a 1–5 scale, where 5 is the highest. The highest weighted score indicates the tasks with the most importance. Comparing the average scores and weighted scores, one can tell that the weights distinguish Tasks 2 and 3 in this example.

The total weight for all factors in a matrix should be equal to 1 (referring to Chapter 3 for the weight selection and normalization). A quick way to do this is for every team member to assign preliminary weights, the team leader to average them for each factor, and then finalize the weight of a factor by dividing by the sum of all factors' weights.

As a simple tool, a prioritization matrix has been used in quality and project fields (Bryson 2018). For example, a prioritization matrix scoring system was used to score medical screening quality indicators (Decker et al. 2019).

**Table 8.10** Sample format of a prioritization matrix.

| Factor | Quality | Function | Time | Budget | ... | Average score | Weighted score |
|--------|---------|----------|------|--------|-----|---------------|----------------|
| Weight | 0.25 | 0.3 | 0.2 | 0.25 | ... | | |
| Task 1 | 5 | 4 | 2 | 4 | | 3.75 | 3.85 |
| Task 2 | 4 | 3 | 3 | 3 | | 3.25 | 3.25 |
| Task 3 | 3 | 5 | 3 | 2 | | 3.25 | 3.35 |
| ... ... | | | | | | | |

### 8.3.7 Process Decision Program Chart

A process decision program chart (PDPC) appears visually similar to a tree diagram. Figure 8.24 shows an example of PDPC for a quality improvement project. If used for goal structuring, a PDPC takes a team from a basic starting point to the final, complex goal. A PDPC chart may include additional information, such as risks.

A PDPC often has at least three levels. The first level is seen as a business or project objective. The second level is main sub-objectives or activities. The third level is about the tasks under each main activity. Importantly, the potential risks and issues associated with activities and tasks should be brainstormed first and included in a PDPC. The fourth level may list these potential issues.

The development of a PDPC could be based on a tree diagram if it is already developed. With tree branches and elements, a team can brainstorm their unique contributions and/or risks. The risks are about what might go wrong in a plan under development or the contribution of an activity to the objective. If no tree diagram is available, one may sketch one and start from there.

For analyzing potential issues, a PDPC is similar to FMEA, as both target these ideas, estimate their risks, and propose contingency actions. A PDPC is relatively easier to comprehend than an FMEA, and is often used for the development of a contingency plan and countermeasures for potential problems.

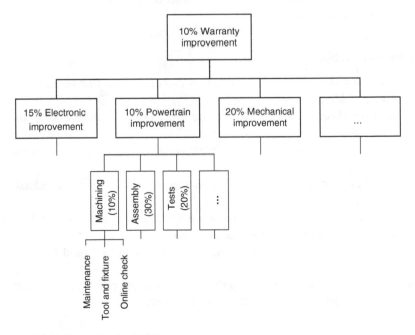

**Figure 8.24** Example of a PDPC.

It may be a good idea to jointly use PDPC and other quality tools, for example, implementing the information gathered from a PDPC into an FMEA. One study has proposed using a PDPC as a strategic decision-making tool for handling the innovation process with considerable rates of failure (Chen and Reyes 2017).

## Summary

### Problem-solving Process

1 CI is a process and principle, which makes a problem-solving process proactive.
2 PDCA is a widely used process for problem solving, and can be used iteratively for CI and problem solving.
3 DMAIC is another widely used process for problem solving.
4 The 8D approach has eight steps for problem solving; and its simplified version is the 5D approach.
5 Problem-solving approaches, e.g. PDCA, DMAIC, 8D, A3, etc., are of a similar process flow, with minor variation in their unique steps and focuses.
6 Three core elements of various problem-solving processes are problem identification, root cause analysis, and solution implementation.
7 For effective problem solving, a given process and appropriate quality tools should be integrated.

### Seven Basic Tools

8 A cause-and-effect (fishbone) diagram can be used to find overall factors for a problem.
9 A check sheet is a simple, structured form, used to collect data through simple data analysis.
10 A histogram chart is used to approximate the frequency distributions of data, using between 5 and 30 bins.
11 A Pareto chart is a bar graph that shows the relative significance of factors, or the frequency of problems.
12 A scatter diagram is a two-dimensional X–Y graphic, showing the relationship between a pair of numerical data sets.
13 Control charts are a family of diagrams of various types, used for process monitoring purposes.
14 Stratification analysis and graphics are used to illustrate sorted data into categories, groups, or layers based on similarity and patterns.

**Seven Additional Tools**

15  An affinity diagram is a mapping tool, used to generate ideas for a specific problem and its factors and linking them to form patterns of thought.

16  A relation diagram is a network diagram to depict causal relationships among multiple items.

17  A tree diagram is used to break a broad goal or item down into a few branches of finer detail. Quantitative data can be used in a tree diagram.

18  A matrix chart (diagram) can show the relationship between two or more groups.

19  A network diagram is used to represent a sequence of activities or tasks in a network, and identify the critical paths of a project.

20  A priority matrix is a structured tool to figure what matters most. Weight factors can be introduced in a priority matrix.

21  A PDPC is used to break down project objectives into sub-objectives or activities, often with three to four levels.

# Exercises

### Review Questions

1  Discuss the mindset of CI.

2  Explain a PDCA process with an example.

3  Which step of PDCA can be challenging, based on your experience or observation?

4  "One suggests that small loops or incremental efforts may make PDCA effective." Do you agree with this statement? Why or why not?

5  Discuss the meaning of a standard update after a PDCA project.

6  Explain a DMAIC process with an example.

7  Explain an 8D process with an example.

8  The D3 objective (to find and implement an interim plan) in the 8D process sounds unique to a problem-solving process. Discuss the necessity of D3 task.

9  Discuss when it is appropriate to use a 5D process instead of an 8D process.

10  Explain a DIVE process with an example.

11  Different problem-solving processes, i.e., PDCA, DMAIC, 8D, 5D, and A3, are similar in many ways. What are their differences (e.g. focus)?

12  Explain a 5W1H process with an example.

13  To brainstorm the potential root causes of a problem, what tools do you suggest using? Why?

**14** To study the root causes of a problem quantitatively, what tools do you suggest using? Why?

**15** If a CI project involves mainly qualitative data, what tools do you suggest using in a PDCA process for the project? Why?

**16** You find multiple factors and potential root causes in a problem-solving project. What tools do you suggest for analyzing and prioritizing these factors and root causes? Why?

**17** Compare a histogram, Pareto chart, and stratification chart for a presentation of data analysis.

**18** Compare a tree diagram, network diagram, and PDPC for quantitatively describing a problem.

**19** For a complex problem, studying the relationships among factors is often a focus. What tools may be used for relationship related studies? Why?

**20** Select one of the 14 quality tools, and discuss its suitable applications.

## Mini-project Topics

**1** Find a case to prove or disprove that "if it ain't broke, don't fix it" still applies today.

**2** Find a complete PDCA case to show how effectively the PDCA process was used.

**3** Compare common problem-solving processes, i.e. PDCA, DMAIC, 8D, and A3, and find their individual advantages and limitations.

**4** Search a research paper on an additional problem-solving approach from library databases, or https://scholar.google.com, and discuss its differences in relation to more common processes.

**5** Compare a pair of quality tools, e.g. fishbone vs. affinity diagram, histogram vs. Pareto chart, or tree diagram vs. stratification. Find their different application targets.

**6** Review the characteristics of the 14 quality tools, and provide a general guideline for their applications in problem solving.

**7** "One states that using a predetermined, proven tool can be effective, but the tool may or may not be the best one for a specific problem." Please debate this statement with a supporting example.

**8** Propose two suggestions how to integrate a given problem-solving process and quality tools for an application.

**9** Review the pros, cons, and applicability of qualitative and quantitative methods in problem solving.

**10** In addition to the 14 tools, is there any other tool that can be used for the same purposes? Discuss one, providing an explanation.

# References

Alpendre, F.T., Batista, J., Gaspari, A. et al. (2017). PDCA cycle for preparing a surgical safety checklist, *Cogitare Enferm*, Vol. 22, Iss. 3, pp.e50964. 10.5380/ce. v22i3.50964.

Aras, Ö. and Özcan, B. (2016). Cost of poor quality in energy sector, *International Journal of Commerce and Finance*, Vol. 2, Iss. 1, pp. 25–35.

Bateni, S., Bold, R., Meyers, F. et al. (2018). Comparison of common risk stratification indices to predict outcomes among stage IV cancer patients with bowel obstruction undergoing surgery, *Journal of Surgical Oncology*, Vol. 117, Iss. 3, pp. 479–487. 10.1002/jso.24866.

Bryson, C. (2018). Prioritization matrix use in program/project management, *Quality*, Vol. 57, Iss. 9, pp. 20–20.

CDC. (2020). COVID-NET: COVID-19-Associated Hospitalization Surveillance Network, Centers for Disease Control and Prevention. https://gis.cdc.gov/grasp/COVIDNet/COVID19_5.html, accessed in June 2020.

Ceberg, C., Benedek, H., and Knöös, T. (2018). Automated treatment planning and quality assurance, *Physica Medica*, Vol. 52, pp. 87. 10.1016/j.ejmp.2018.06.305.

Chen, C. and Reyes, L. (2017). A quality management approach to guide the executive management team through the product/service innovation process, *Total Quality Management & Business Excellence*, Vol. 28, Iss. 9–10, pp. 1003–1022. 10.1080/14783363.2017.1303878.

Constantinescu, G.C. (2017). Sources of supply chain conflicts–A fishbone diagram correlation, *SEA: Practical application of science, Editorial Department*, Vol. 5, Iss. 13, pp. 191–197.

Decker, K., Baines, N., Muzyka, C. et al. (2019). Measuring colposcopy quality in Canada: Development of population-based indicators, *Current Oncology*, Vol. 26, Iss. 3, pp. e286–291. 10.3747/co.26.4709.

Faloudah, A., Qasim, S. and Bahumayd, M. (2015). Total quality management in healthcare, *International Journal of Computer Applications*, Vol. 120, Iss. 12, pp. 22–24.

Fritsche, P. (2016). New technology, small footprint ($9' \times 4'$), vertical circuit washing system for PCBAS ($< 3" \times 3"$), *2016 Pan Pacific Microelectronics Symposium*, Big Island, HI, pp. 1–7. 10.1109/PanPacific.2016.7428405.

Hanawa, H. and Momo, K. (2019). PDCA cycle for the development of clinical formulation thinking in actual example, *Journal of the Pharmaceutical Society of Japan*, Vol. 139, Iss. 10, pp. 1267–1268. 10.1248/yakushi.19-00121-F.

Hughes, C., (2015). Unit Exercise, Eastern Michigan University QUAL 548 Tools for Continuous Improvement (CRN 16793), November 23, 2015.

Jin, H., Huang, H., Dong, W. et al. (2012). Preliminary experience of a PDCA-cycle and quality management based training curriculum for rat liver transplantation,

*Journal of Surgical Research*, Vol. 176, Iss. 2, pp. 409–422. 10.1016/j. jss.2011.10.010.

Júnior, A. and Broday, E. (2019). Adopting PDCA to loss reduction: A case study in a food industry in Southern Brazil, *International Journal for Quality Research*, Vol. 13, Iss. 2, pp. 335–348. 10.24874/IJQR13.02-06.

Lenort, R., Staš, D., Holman, D. and Wicher, P. (2017). A3 method as a powerful tool for searching and implementing green innovations in an industrial company transport, *Procedia Engineering*, Vol. 192, pp. 533–538. 10.1016/j. proeng.2017.06.092.

Li, R., Wang, W., Xu, X. et al. (2018). Cooperative planning model of renewable energy sources and energy storage units in active distribution systems: A bi-level model and Pareto analysis, *Energy*, Vol. 168, pp. 30–42. 10.1016/j. energy.2018.11.069.

Martin, G. (n.d.). The meaning and origin of the expression: If it ain't broke, don't fix it, https://www.phrases.org.uk/meanings/if-it-aint-broke-dont-fix-it.html, accessed in August 2020.

Memon, I.A., Jamali, Q.B., Jamali, A.S. et al. (2019). Defect reduction with the use of seven quality control tools for productivity improvement at an automobile company, *Engineering, Technology & Applied Science Research*, Vol. 9, Iss. 2, pp. 4044–4047.

Roy, A., Cutright, D., Gopalakrishnan, M. et al. (2019). A risk-adjusted control chart to evaluate intensity modulated radiation therapy plan quality, *Advances in Radiation Oncology*, 10.1016/j.adro.2019.11.006.

Sari, D.K., Hetharia, D., Saraswati, D. and Marizka, R. (2019). Design of flat shoes quality control system using PDCA (case study at PT DAT), *IOP Conference Series: Materials Science and Engineering*, Vol. 528, 012073. 10.1088/1757-899X/528/1/012073.

Schmidt, M. (2012). General Motors Technical Problem Solving Group Drives Excellence, *ASQ Case Study*, http://asq.org/public/wqm/general-motors.pdf, accessed in May 2020.

Schrage, M. (2017). AI is going to change the 80/20 rule, *Harvard Business Review*, https://hbr.org/2017/02/ai-is-going-to-change-the-8020-rule, accessed in September 2020.

Sobek, D.K. and Jimmerson, C. (2004). A3 Reports: A tool for process improvement and organizational transformation, *Proceedings of the Industrial Engineering Research Conference*, Houston, Texas. May 16–18, 2004.

Tang, H., (2018). Manufacturing System and Process Development for Vehicle Assembly. Warrendale, PA: SAE International. https://www.sae.org/publications/books/content/r-457.

Tang, H. (2021). *Engineering Research – Design, Methods, and Publication*. Hoboken, NJ: Wiley.

Tenbergen, B., Weyer, T., and Pohl, K. (2018). Hazard relation diagrams: A diagrammatic representation to increase validation objectivity of requirements-based hazard mitigations, *Requirements Engineering*, Vol. 23, pp. 291–329. 10.1007/s00766-017-0267-9.

Xu, H., Lu, J., Wang, J. et al. (2019). Implement a knowledge-based automated dose volume histogram prediction module in Pinnacle3 treatment planning system for plan quality assurance and guidance, *Journal of Applied Clinical Medical Physics*, 10.1002/acm2.12689

# Acronyms and Glossary

(A brief reference for the items used without elaborate discussion in this book)

**ACSI**  American Customer Satisfaction Index, an indicator of the consumer satisfaction with the goods and services across the US economy sectors, based on surveys and interviews.

**AHRQ**  The Agency for Healthcare Research and Quality, an agency within the US Department of Health and Human Services.

**AIAG**  Automotive Industry Action Group, an automotive professional, in quality related areas.

**AS 9100D**  A standard for requirements for the quality management systems of aviation, aerospace, and defense industries. The first version was released in 1999.

**AS 9145**  An aerospace standard for the requirements for advanced product quality planning and production part approval process.

**ASTM**  An international standards organization for a wide range of materials, products, systems, and services, formerly known as American Society for Testing and Materials.

**ASTM E2782**  a standard guide for measurement systems analysis (MSA).

**CDC**  The Centers for Disease Control and Prevention, the national public health agency of the United States.

**CFR**  The Code of Federal Regulations, the US government's general and permanent rules and regulations.

**CMM**  Coordinate measuring machine, a device to measure the geometry of a physical object by sensing its surface using a probe.

**DOD**  The United States Department of Defense

**DOE**  The United States Department of Energy

**Downtime**  A status or duration, out of operation or unavailable for use, of equipment, machine, or system.

*Quality Planning and Assurance: Principles, Approaches, and Methods for Product and Service Development*, First Edition. Herman Tang.
© 2022 John Wiley & Sons, Inc. Published 2022 by John Wiley & Sons, Inc.

**Durability** Ability of a product working in an environment without requiring excessive maintenance or repair.

**ECSI** European Customer Satisfaction Index, similar to ACSI.

**EPA** The Environmental Protection Agency, an agency of the United States federal government.

**FTQ or FTTQ** First-time Quality or First-Time-Through Quality, a quality performance indicator of a system. Normally presented as a percentage of units produced without quality issues.

**HBR** Harvard Business Review, a general management magazine, published by Harvard Business Publishing, a wholly owned subsidiary of Harvard University.

**IATF** International Automotive Task Force, a group of automotive manufacturers, formed to provide improved quality products to automotive customers worldwide.

**IATF 16949** An international technical specification for the quality management systems in the automotive sector.

**IEC 61025** An international standard for fault tree analysis and guidance on its applications

**IEEE** Institute of Electrical and Electronics Engineers, a professional association for electronic engineering and electrical engineering.

**ISO 10012** An international standard as the generic requirements and guidance for the management of measurement processes and metrological confirmation of measuring equipment.

**ISO 12132** International guidelines for preparing design failure mode and effects analysis (FMEA).

**ISO 13485** An international standard for quality management system for medical devices.

**ISO 14001** International standards for the criteria for an environmental management system.

**ISO 16355** International standards for quality function deployment (QFD) process, its purpose, users, and tools.

**ISO 22000** An international standard for food safety management systems.

**ISO 31000** An international standard for risk management.

**ISO 9000** A set of international standards for quality management systems.

**ISO 9001** International standards for quality management systems.

**ISO/IEC 17025** An international standard for general requirements for the competence of testing and calibration laboratories.

**ISO/IEC 20000** An international standard for IT service management.

**ISO/IEC 27001** An international standard for requirements for an information security management system (ISMS).

**J.D. Power**  A data analytics, market research, and consumer intelligence company.

**Kappa (κ) statistic**  A statistic to measure inter-rater reliability for qualitative items, considering the possibility of the agreement occurring by chance.

**Kendall's statistic**  A statistic used to measure the ordinal association between two measured quantities.

**Likert scale**  A scale commonly involved in survey research that employs questionnaires.

**Maintainability**  The ability of an equipment, machine, or product, to be retained in or restored to a specified condition when maintenance is performed.

**MTBF**  Mean Time between Failures, a reliability indicator of a product or process.

**MIL–HDBK–338B**  A military handbook: electronic reliability design handbook.

**Minitab**  A statistical data analysis package.

**Mockup**  A model or replica of a machine or structure, used for test or experimental purposes.

**Mystery shopping**  an assessment method used for marketing research, by independent contractors.

**NASA**  The National Aeronautics and Space Administration, an agency of the US federal government.

**NDA**  Non-disclosure agreement, a legally enforceable contract on the confidentiality between two parties.

**NIH**  The National Institutes of Health, an agency of the United States federal government.

**NUREG–0492**  Fault Tree Handbook, by the US Nuclear Regulatory Commission.

**OEE**  Overall equipment effectiveness, an indicator of how well a manufacturing operation compared to its full potential, during a period.

**PLM**  Product lifecycle management, a handling system of a good through the typical stages of its product life.

**PO**  Purchase Order, an official document to a seller indicating types, quantities, and agreed prices for a product or service.

**Q1**  A certification to Ford's suppliers, with excellence beyond the ISO/TS certification requirements.

**QAPP**  Quality Assurance Program Plan, the procedures and requirements to monitor a project for government and service sectors, similar to the APQP for manufacturing industries.

**Reliability**  Probability of a product or service working consistently well without major failure.

**SAE**  A professional global association in the aerospace, automotive and commercial-vehicle industries, previously known as the Society of Automotive Engineers.

**SAE ARP4761**  Industry guidelines and methods for conducting the safety assessment process on civil airborne systems and equipment is an aerospace recommended practice.

**SAE J1739**  An industry standard for potential FMEA, including design FMEA, supplemental FMEA-MSR, and process FMEA.

**SAE J2886**  An industry guideline for Design Review Based on Failure Modes (DRBFM) recommended practice.

**SCSB**  Swedish Customer Satisfaction Barometer, an original development of ACSI.

**Serviceability**  Quality of being able to provide defined service or usage.

**TPS**  Toyota Production System, a management system of integrated manufacturing and logistics, and interaction with suppliers and customers for Toyota.

**TQM**  Total Quality Management, a management approach to integrate all employees and functions, and use various quality tools for customer-orientated quality.

**VDA**  German Association of the Automotive Industry, including automobile manufactures and their component suppliers. A member of the European Automobile Manufacturers Association.

**V-shape model**  A graphical representation of a systems engineering process and development lifecycle of a product or service, emphasizing requirement-driven design and testing.

# Epilogue

As reviewed in Chapter 1, the Juran Trilogy of quality management systems (QMS) comprises three main pillars: 1) quality planning, 2) quality control, and 3) continuous improvement. Among these three pillars, quality control is conventional, and continuous improvement is an active practice in industries. However, as the proactive pillar, quality planning often is not given enough attention in practice. In this book, I have concisely presented most aspects of quality planning from a quality system standpoint.

There are numerous opportunities in the development and implementation of the principles and approaches to quality planning. In the real world, every situation can be different, so one would need to be adaptive when applying quality planning principles and approaches. In teaching my graduate quality planning class, I encourage everyone to try something creative for each application, based on the principles and approaches learned. Through innovative quality planning, quality professionals can set a high level of product and service quality going forward in the future.

I wish readers continued success and enjoyment on their quality management journey!

*Quality Planning and Assurance: Principles, Approaches, and Methods for Product and Service Development*, First Edition. Herman Tang.
© 2022 John Wiley & Sons, Inc. Published 2022 by John Wiley & Sons, Inc.

# Index

*Quality Planning and Assurance: Principles, Approaches, and Methods for Product and Service Development*, First Edition. Herman Tang.
© 2022 John Wiley & Sons, Inc. Published 2022 by John Wiley & Sons, Inc.

9 781119 819271